40 DAYS and 40 NIGHTS

★ DARWIN, *Intelligent Design*,

GOD, *OxyContin,®* AND OTHER *Oddities*

ON TRIAL IN *Pennsylvania* ★

40 DAYS *and* 40 NIGHTS

MATTHEW CHAPMAN

Collins

An Imprint of HarperCollins*Publishers*

40 DAYS AND 40 NIGHTS. Copyright © 2007 by Matthew Chapman. All rights reserved. Printed in the United States of America. No part of this book may be used or reproduced in any manner whatsoever without written permission except in the case of brief quotations embodied in critical articles and reviews. For information, address HarperCollins Publishers, 10 East 53rd Street, New York, NY 10022.

HarperCollins books may be purchased for educational, business, or sales promotional use. For information, please write: Special Markets Department, HarperCollins Publishers, 10 East 53rd Street, New York, NY 10022.

First Collins paperback edition published 2008

Designed by Ellen Cipriano

The Library of Congress has catalogued the hardcover edition as follows:

Library of Congress Cataloging-in-Publication Data

Chapman, Matthew, 1950–
40 days and 40 nights : Darwin, intelligent design, God, Oxycontin®, and other oddities on trial in Pennsylvania / Matthew Chapman.
p. cm.
ISBN 978-0-06-117945-7
1. Kitzmiller, Tammy—Trials, litigation, etc. 2. Dover Area School District (Dover, Pa.)—Trials, litigation, etc. 3. Trials—Pennsylvania—Dover. 4. Evolution (Biology)—Study and teaching—Law and legislation— Pennsylvania—Dover. 5. Scopes, John Thomas—Trials, litigation, etc. I. Title. II. Title: Forty days and forty nights.

KF228.K589 2007
345.748'0288—dc22 2006049738

ISBN 978-0-06-117946-4 (pbk.)

08 09 10 11 12 ID/RRD 10 9 8 7 6 5 4 3 2 1

This book is dedicated to my family and other animals, and to all those who believe, like George Santayana, that "Scepticism is the chastity of the intellect, and it is shameful to surrender it too soon."

"I will cause it to rain upon the earth forty days and forty nights, and every living substance that I have made will I destroy from off the face of the earth."

God, speaking in Genesis

"By my reckoning, this is the fortieth day since trial began and tonight will be the fortieth night, and I would like to know if you did that on purpose?"

Patrick Gillen, attorney in *Kitzmiller v. Dover*, speaking to the judge

"Mr. Gillen, that is an interesting coincidence, but it was not by design."

Judge John Jones

CONTENTS

Dramatis Personae xi

PART ONE ★ In the Beginning 1

Genesis 3
Early Days 14
Bill and Alan 24
Life on Board 29
Miller Grinds It Up 34
Other Primates 52
The Teachers' Story 58
The Audible Ping 70
A Libertarian Buddhist Girl Scout Leader 78
Pennock qua Pennock, Finches qua Finches 84
Summertime 90
Marilyn Monroe Is Alive and Well and Living in Dover 99

John Haught and the Teapot of Wisdom 104
The Panda Cometh 118
The Smoking Gun 130
Barbara Forrest and the Panda's Tale 138
When You Open Your Eyes in Hell 150
Kevin Fills the Gaps 156
You Pay Your Nickel and You Go for a Ride 161

PART TWO ★ The New Testament of Science 165

The Man from Bethlehem 167
Moral Decay 198
Sudden Disappearance, Sudden Emergence 206
Buckingham 219
The Reporters 227
Bonsell and His Trinity of Loyal Women 234
The Two Scotts 243
The Genie Is Out of the Bottle 253
Revelation 263

DRAMATIS PERSONAE

PLAINTIFFS

TAMMY KITZMILLER—Lead plaintiff, mother of two girls in Dover High.

ARALENE "BARRIE" CALLAHAN—Ex–school board member, mother of three children, one in Dover High.

FRED CALLAHAN—Local businessman, husband of Barrie Callahan.

BRYAN REHM—Science teacher at Dover High.

CHRISTIE REHM—Teacher at a local school, Bryan Rehm's wife.

BETH EVELAND—Legal assistant to a local attorney, mother of two.

CINDY SNEATH—Co-owner of an appliance repair shop, Tammy Kitzmiller's neighbor, mother of two.

JULIE SMITH—Medical technologist, divorced mother of two, one daughter in tenth grade at Dover High.

STEVE STOUGH—Teacher and coach, parent of a daughter in eighth grade at Dover High.

JOEL LIEB—Teacher, family in Dover since its earliest beginnings.

DEB FENIMORE—Lived with Joel Lieb, worked at a youth advocacy program, mother of a child by Joel.

SCHOOL BOARD MEMBERS

ALAN BONSELL—Owner of a local auto repair shop, property owner/developer, Protestant fundamentalist, pro–intelligent design and creationism.

BILL BUCKINGHAM—Retired cop and corrections officer, Protestant fundamentalist, pro–intelligent design and creationism.

SHEILA HARKINS—Property owner/developer, Quaker, pro–intelligent design.

HEATHER GEESEY—Wife and full-time mother, pro–intelligent design.

JANE CLEAVER—Retired owner of a five-and-dime, pro–intelligent design.

ANGIE YINGLING—Owner of a local auto repair shop, property owner/developer, Marilyn Monroe fan, had shifting views.

NOEL WENRICH—Creationist, eventually opposed intelligent design.

JEFF BROWN—Electrician, Sunday School teacher, Protestant, anti–intelligent design.

CAROL "CASEY" BROWN—Former reporter for local paper, degree in education, wife of Jeff Brown, Protestant, anti–intelligent design.

LAWYERS FOR THE PLAINTIFFS

ERIC ROTHSCHILD—Lead attorney for the plaintiffs, corporate litigator at Pepper Hamilton in Philadelphia.

STEVE HARVEY—Eric Rothschild's co-counsel, also of Pepper Hamilton.

WITOLD "VIC" WALCZAK—Head of the Pennsylvania ACLU, based in Pittsburgh.

RICHARD KATSKEE—Lawyer from Americans United for Separation of Church and State.

THOMAS SCHMIDT—Pepper Hamilton attorney.

LAWYERS FOR THE DEFENSE

RICHARD THOMPSON—Head of the Thomas More Law Center, which represented the Dover Area School District; attorney for the defense.

PATRICK GILLEN—Lead attorney for the defense.

ROBERT MUISE—Attorney for the defense.

EDWARD WHITE III—Attorney for the defense.

EXPERT WITNESSES

For the Plaintiffs

KEN MILLER—Professor of biology at Brown University, co-author of *Biology*.

ROBERT PENNOCK—Professor at Michigan State University, degrees in biology and philosophy.

JOHN HAUGHT—Catholic theologian, recently retired from chairing the theology department at Georgetown University.

BARBARA FORREST—Professor of philosophy at Southeastern Louisiana University, author of *Creationism's Trojan Horse: The Wedge of Intelligent Design*.

KEVIN PADIAN—Paleontologist, professor of integrative biology at the University of California in Berkeley, curator in the Museum of Paleontology, president of the National Center for Science Education.

BRIAN ALTERS—Professor of science education at McGill University.

For the Defense

MICHAEL BEHE—Biologist, author of *Darwin's Black Box*, proponent of intelligent design, and the man who allegedly coined the expression "irreducible complexity."

STEPHEN FULLER—Professor of sociology at the University of Warwick, England.

SCOTT MINNICH—Professor of microbiology at the University of Idaho.

OTHER PLAYERS

HEDYA ARYANI—Legal assistant from Pepper Hamilton.

MICHAEL BAKSA—Dover assistant school superintendent.

HEIDI BERNARD-BUBB—Part-time local reporter. Covered early school board meetings.

WILLIAM DEMBSKI—One of the earliest fellows of the Discovery Institute's Center for Science and Culture.

ROBERT ESHBACH—Biology teacher at Dover High.

KATE HENSON—Legal assistant from Pepper Hamilton.

PAULA KNUDSEN—Staff attorney at the ACLU in Harrisburg.

ROBERT LINKER—Science teacher at Dover High.

JOE MALDONADO—Part-time local reporter. Covered early school board meetings.

NICK MATZKE—Staffer at the National Center for Science Education and scientific adviser to the plaintiffs' legal team.

MATTHEW MCELVENNY—Technology specialist for the plaintiffs' legal team.

STEPHEN MEYER—Founder of the Discovery Institute's Center for Science and Culture.

JEN MILLER—Biology teacher at Dover High.

RICHARD NILSEN—Dover school superintendent.

EUGENIE SCOTT—Executive director of the National Center for Science Education, an organization that defends against attacks on evolution in schools.

BERTHA "BERT" SPAHR—Head of the science department at Dover High, chemistry teacher at the school for forty-one years.

SCOPES TRIAL CHARACTERS

JOHN SCOPES—School teacher in Dayton, Tennessee; put on trial for teaching evolution in 1925.

CLARENCE DARROW—Defense lawyer for Scopes.

DUDLEY MALONE—Defense lawyer for Scopes.

WILLIAM JENNINGS BRYAN—Fundamentalist ex–Democratic Presidential candidate, on the team prosecuting Scopes.

THE JUDGE

JOHN JONES III

PART I

In the BEGINNING

Genesis

I HOPE THE FACT that my great-great-grandfather was Charles Darwin will not deter you from reading this book. You might assume that my opinions are predictable and that a less biased, and therefore more suspenseful, account could be found elsewhere. The truth is that at the start of the trial I did believe creationism should be banned from high school science classes. By the end of it, however, I had been convinced by the intelligent design advocates that creationism in all its forms should be a mandatory part of every child's science education. My reasons for believing this are slightly different from theirs, but that's another story—the story of this book.

. . . .

Being a descendent of Charles Darwin was not something I thought much about as I was growing up in Cambridge, England. The theory of evolution was accepted, and Darwin was a mere historical figure. If I did think about my connection to him, it was only negatively. Academic pressure on me was intense, and, at least in comparison with my ancestor, success was unlikely.

I was a child whose maximum attention span was approximately five seconds, a boy who refused to be educated and was kicked out of several schools, and a youth whose only academic achievement was a stunning lack of achievement. At the age of fifteen, when I was set free, I had not passed a single exam of any consequence. Soon after the school door slammed behind me, I rediscovered my curiosity.

To support myself, I worked in a variety of jobs—van driver, welder, house cleaner, bricklayer, spotlight operator in a nightclub, and so on—before becoming an apprentice film editor, editor, screenwriter, and finally film director.

In the early eighties, I moved to the United States, where I discovered to my surprise that many Americans not only rejected Darwin's theory of evolution, but they reviled it. I had come here in part because I hated the English class system and thought of America as being less weighed down by the past. Now here I was in the New World, faced with an old and willful ignorance that went far beyond anything even I had attempted.

I did not know much about evolution, but a quick study of easily available information showed that its most important idea, natural selection, was easy to understand and made sense.

Darwin saw how plant and animal breeders influenced characteristics through selective breeding. Why wouldn't nature do the same? If life was a struggle for survival, those best suited to their environment had an advantage. Any small, random mutation favoring survival would increase the likelihood of that animal or plant living long enough to pass on its genes to offspring, who would then inherit the advantage, and so on. Increased complexity and slow adaptation seemed inevitable.

It soon became apparent from my reading that 99 percent of scientists believed in evolution. Why would one doubt them? Did the pedestrian question the theory of gravity? Did the farmer who went to the doctor question his diagnosis? Why, when it came to evolution, did nonexperts feel compelled to disagree with those who clearly knew better?

The answer was that evolution appeared to contradict the bible. Evolution requires a lot of time to bring about change, and if plants and animals constantly become more complex, it was logical to infer that previously they had been far simpler. If one went back far enough, it seemed probable, though hard to prove, that all life-forms on earth shared a primitive ancestor perhaps found in some distant "primeval soup" of chemicals.

This, of course, was not how the origin of life was described in the bible. Evolution did not put either God or human beings at the center of a recent, ordered, and purposeful creation but instead suggested a long, brutal, and random process.

By the time I arrived in America, the evidence for the basic ideas of evolution had been overwhelming for a century. However, given a choice between it and the more comforting biblical version, most people chose the latter.

This was the beginning of my interest in irrational beliefs of all kinds. Why did so many people in an otherwise confident and sophisticated country cling to their faith in so many things—from astrology to the Zohar—that were so often contradicted by evidence and reason?

In 2001, I wrote *Trials of the Monkey: An Accidental Memoir* about my childhood and early working life interspersed with an account of the so-called Scopes Monkey Trial.

In 1925, schoolteacher John Scopes was prosecuted for teaching evolution in a Tennessee high school, contravening a recent state law, the Butler Act. The first trial ever to be broadcast live—by radio—not only to America but also to Europe and Australia, it was a fantastic philosophical skirmish between religion and reason, between the most famous fundamentalist of his day, William Jennings Bryan, and the most famous humanist, Clarence Darrow. Played out in a circus-like atmosphere, it became the basis for the film *Inherit the Wind*.

I visited Dayton, Tennessee, where the trial took place, and fell in love with both the town and the trial, with its hilarious mix of philosophy and hucksterism. In my mind the antievolution movement remained a quaint Southern aberration resulting from a combination of moonshine and religions of the snake-fondling type. I had drunk some of the aforementioned mountain dew and found it a powerful mind-altering substance, oddly delicious, with only the faintest leady aftertaste of the car radiator through which it had been distilled, but concluded it was not the best stimulant of intellectual cogitation.

Early in 2005, I began to read reports of a Scopes-like case developing in Dover, Pennsylvania, but did not believe such a thing could reach fruition only three and a half hours from New York City, and if it did, I could not imagine it providing anything like as much amusement.

It soon became apparent, however, that *Kitzmiller v. Dover*, which would be tried in federal court in Harrisburg, though not attracting as much attention as Scopes, had something the earlier trial lacked.

Clarence Darrow's expert witnesses, a number of eminent scientists, were excluded by the judge, so William Jennings Bryan, a creationist, attacked evolution without contradiction. Here he is reading from Darwin's *The Descent of Man* and then commenting on it afterwards.

> We may thus ascend to the Leuridae, and the interval is not very wide from these to the Simiadae. The Simiadae then branched off into two great stems, the new world and the old world monkeys and from the latter, at a remote period, man, the wonder and glory of the universe, proceeded." Not even from American monkeys, but from old world monkeys!

He got, according to the court record, a pretty good laugh for this, and then continued:

> They come here with this bunch of stuff that they call evolution, that they tell you that everybody believes in, but . . . they do not explain the great riddle of the universe—they do not deal with the problems of life—they do not teach the great science of how to live—and yet they would undermine the faith of these little children in God who stands back of everything and whose promise we have that we shall live with Him forever by and by.

Robbed of his scientific testimony, Darrow retaliated by challenging Bryan to take the stand as an expert witness on the bible, then asked him a series of questions designed to show that the good book was perhaps not the best document from which to teach science.

Did Bryan really believe that the earth was created in six literal days less than ten thousand years ago? Where did Cain get his wife? If Joshua commanded the sun to stand still in the sky, did this mean that at that time the sun went around the earth? And so on.

Highly amusing, not without purpose, but hardly the highest level of debate on the validity of either evolution or religion. In *Kitzmiller v. Dover,* however, everyone was welcome. Intelligent design claimed to be a scientific theory, so its scientific advocates—the leading one a biologist—could hardly ask that other scientists be excluded.

Intelligent design, or I.D. as it is often known, though complicated in its details, is actually a simple idea that has been around a long time. It proposes that some things in nature appear so complex that they could not have been formed by the slow successive steps required by evolution. They must therefore have been designed in their existing form by an "intelligent designer." The more biologically complex an organism is, the more likely it is that it was designed.

Biologists on both sides would get top billing, but there would also be paleontologists, philosophers, and even theologians. And that wasn't all: the locals would take the stand, too, those for evolution and those against.

A bomb had exploded in a small town and blown it apart. The autopsy was about to begin. Here was a trial, I told myself grandly, that could reveal America's soul—if you believed in souls. Having no idea how many of my preconceptions about Americans were about to be challenged, nor any premonition that I was about to meet the best and worst of them—and that I would, curiously, like most of them—I packed my bags.

• • • •

Many people I met in Pennsylvania believed that all of America was settled by people whose sole motivation was to create a country where democracy and religious tolerance prevailed. In truth, the men who colonized Virginia were there to make a buck, while the Pilgrims of early Massachusetts often denied non-Puritans the right to vote and were always unwelcoming to witches. But if any state *was* founded for almost entirely idealistic reasons, it was in fact their own state, the state of Pennsylvania.

Founded by William Penn in the late 1600s, its initial purpose was to provide sanctuary for Penn's fellow Quaker brethren, but people of all faiths were soon invited to participate in a Holy Experiment, a brand-new concept: the idea that everyone should have the "right to worship Almighty God according to the dictates of their own conscience."

Though Penn was essentially a pacifist, York County, in which Dover lies, soon gained a reputation for military enthusiasm, supplying, for example, more troops for the Revolution than any other area of comparable size and population. The tradition continued, as was evidenced by the funerals of local boys killed in Iraq.

With *Kitzmiller v. Dover,* another paradox was added. The site of Penn's Holy Experiment became known for a narrow religious zealotry that seemed in total contradiction to Penn's expansive hopes and dreams.

Modern Dover is a small rural town whose population, including a recent influx of people who commute to other larger towns, is approaching two thousand. Flanked by a pair of highways descending out of Harrisburg—one heading to York, then on to Baltimore, the other going to Gettysburg, then on to Frederick, Maryland—Dover itself is on a road to nowhere and is consequently hard to find; and when you arrive, you don't know where to stop. There is no logical center, no place where you might say, "Ah, I'm in Dover now; here's the Town Hall, or here's the Market Square." There is a point at which two roads cross, but it is only that: a crossroad with a traffic light.

On the larger of these two roads, the one coming from York, are gathered all the usual shops and services: fast-food restaurants, realtors, a few antique shops, a couple of car dealerships, and a few old diners now depressingly overwhelmed by the fast-food chains. The fire station is near the nonexistent center, the police station is not, and the only bar I ever heard about, the Racehorse Tavern, is several miles away. There are many churches, and then there is Dover High.

Although the population of the town is only two thousand, the Dover Area School District encompasses a wider gene pool, extracting its victims from a predominantly rural population of forty thousand scattered across the surrounding countryside. Consequently, Dover Senior High has about a thousand students, and their arrival and departure in fleets of yellow buses marks the liveliest hours of the town.

The school presents a modern face to the world, a light-colored brick building with a large metallic sign over the main entrance. To one side is a football field. Across the road is a small pizza joint where the kids congregate after school. I later learned that a prayer group met each morning and joined hands around the flagpole, and that many kids walked around inside the school carrying bibles—it was in fact quite fashionable to do so—but they must have stashed them when they saw my cynical figure lurking on the sidewalk because I never witnessed the phenomenon.

In spite of the school's almost lavish exterior, the district was, in fact, constantly broke. Many of the industries that once augmented the agricultural economy had closed down or moved to countries where labor was cheaper. The only large factory to which you could easily commute was over in York. Perhaps it too wanted to outsource, but who would buy a Chinese Harley-Davidson?

In 2004, a war broke out in Dover that made enemies of old friends and friends of people who would never otherwise have known one another. There were tears, screaming matches, political maneuvers, threats, lies, and many inflammatory religious statements that would have horrified William Penn.

"It's driven a wedge where there hasn't been a wedge before," said one Dover resident, whose family had lived in the town since the beginning. "People are afraid to talk to people . . . They're afraid to talk to me because I'm on the wrong side of the fence."

The fence that divided Dover was driven into the ground by fundamentalists on the school board. On their side were all those who, doubting the theory of evolution, wanted first creationism, then its college-educated offspring, intelligent design, taught in the ninth-grade science class. On the other side were all those who believed that under its academic cloak, intelligent design was just a new form of creationism and therefore not only destructive to their children's education but also a violation of the U.S. Constitution.

At one school board meeting, the head of the curriculum committee stood up and shouted, "Two thousand years ago, someone died on a cross. Can't someone take a stand for him?" At another, this same man, an ex-cop and corrections officer, declared that the current biology book was "laced with Darwinism" and that it was "inexcusable to have a book that says man descends from apes with nothing to counterbalance it."

Eventually, he found a book he liked, an intelligent design biology book called *Of Pandas and People*.

Before long the town was split over whether or not the potentially religious *Of Pandas and People* should be included in the science curriculum. According to local newspaper polls, the majority favored the book, but the science teachers at the school hated it, as did many parents and some of the school board members.

The battles became so painful that several board members resigned, and the composition of the board changed. Soon it was dominated by fundamentalists and fellow travelers. After several months, and after much legal advice, both good and bad, the school board came up with what they thought was a compromise, one that would keep their opponents at bay and avoid a costly lawsuit.

Of Pandas and People would not be directly taught in science class. It would, however, be allowed into school as a reference book, and before anyone was taught evolution, the following immunizing statement would be read in class:

> The Pennsylvania Academic Standards require students to learn about Darwin's Theory of Evolution and eventually to take a standardized test of which evolution is a part.
>
> Because Darwin's Theory is a theory, it continues to be tested as new evidence is discovered. The Theory is not a fact. Gaps in the Theory exist for which there is no evidence. A theory is defined as a well-tested explanation that unifies a broad range of observations.
>
> Intelligent Design is an explanation of the origin of life that differs from Darwin's view. The reference book, *Of Pandas and People*, is available for students who might be interested in gaining an understanding of what Intelligent Design actually involves.
>
> With respect to any theory, students are encouraged to keep an open mind. The school leaves the discussion of the Origins of Life to individual students and their families. As a Standards-driven district, class instruction focuses upon preparing students to achieve proficiency on Standards-based assessments.

If this compromise satisfied the board, it did not satisfy several parents in Dover.

Everyone—except perhaps the board—could see a lawsuit coming. An editorial in the *York Daily Record* said, "Watching what's going on in the Dover Area School District is like watching a train wreck in slow motion."

. . . .

Eric Rothschild, the plaintiffs' lead attorney, was a successful lawyer, but his interests and ambitions extended beyond the corporate law from which he earned his keep. Several years before *Kitzmiller*, he had put his name down on a list of attorneys compiled by the Oakland-based National Center for Science Education, which defends against attacks on evolution in schools across America.

In the fall of 2004, he received a general e-mail from Eugenie Scott, the executive director of the NCSE, about some parents in Dover who were thinking of suing their school board. The American Civil Liberties Union was already interested but didn't have enough lawyers or resources to pursue the case alone. Were there any attorneys working with large firms who might want to join in on a pro bono basis?

Rothschild called Eugenie Scott immediately, then called Vic Walczak, head of the Pennsylvania ACLU, which had its head office in Pittsburgh. They got along well on the phone, so Rothschild went to talk to Steve Harvey, a colleague of his with whom he had worked several times before. Harvey was interested. Next, Rothschild went to visit the attorney at his law firm, Pepper Hamilton in Philadelphia, who decided which pro bono cases the firm took on. Pepper Hamilton had a history of doing pro bono work. They had, for example, represented the reproductive rights side in *Planned Parenthood v. Casey*, which was the closest *Roe v. Wade* ever came to being reversed. The attorney agreed that the case was interesting and allowed Rothschild to proceed.

Within a matter of days, the deal between the three legal entities that would represent the plaintiffs—the ACLU, Americans United for Church and State, and Rothschild and Harvey for Pepper Hamilton—was worked out. As he intended to do more work than anyone else, Rothschild insisted on being lead counsel.

Eventually, he and his colleagues gathered eleven parents who wanted to sue the school board in federal court. They would argue that students were being deprived of their rights under the Establishment Clause of the First Amendment, which prohibits—according to most interpretations—the teaching or presentation of religious ideas in public school science classes.

At a press conference in Harrisburg on December 14, 2004, the plaintiffs announced their lawsuit. Unless the school board changed its policy, they would be seeking "declaratory and injunctive relief, nominal damages, costs, and attorney fees." To put this in lay terms, this meant the parents wanted intelligent design out of the high school—and if they won, the school board would get the bill. On top of Rothschild and Harvey's fees and Pepper Hamilton's costs were those of the ACLU and Americans United. If the impoverished school district lost the case, the financial penalty could easily run into the millions.

The school board did not blink. Instead, its members accepted the offer of a Catholic law firm out of Michigan to defend them.

Ten months later, the case began.

Early Days

THE COMFORT INN, where I stayed during the six weeks of *Kitzmiller v. Dover*, is in downtown Harrisburg. It overlooks the Susquehanna River and a series of beautiful bridges that cross it. A cooling breeze blew off the river but never entered the hotel. The windows were sealed shut. The Comfort Inn was its own biosphere. Your climatic choices were limited to Fan, Low Heat, High Heat, Low Cool, High Cool, and Off.

The morning after I arrived, I went for a run alongside the river. As I had to be in court by 8:30 to pick up my press pass for the opening of court at 9:00, it was still dark, but I have a tendency to claustrophobia and hypochondria and wanted to give my lungs a few brief gasps of unfiltered air. As I ran I contemplated some research I had done. The National Academy of Sciences, perhaps the leading scientific organization in the world, had recently stated that "It is no longer possible to sustain the view that living things did not evolve from earlier forms or that human beings were not produced by the same evolutionary mechanisms that apply to the rest of the living world."

And yet, in spite of mounting evidence confirmed almost on a weekly basis by genetics, biochemistry, and the fossil record, a 2005 poll found that 54 percent of adults in the United States did not think human beings evolved from earlier species. This was up from 46 percent in 1994. Even the president, George W. Bush, believed "the jury was still out" on evolution.

Creationists talk about "gaps" in Darwin's theory. The above statistics represented a far more interesting gap. It was the gap between evidence and belief, between rationality and faith.

After breakfast I walked up through town to the federal courthouse, an unremarkable government building on Walnut Street. News cameras and many reporters already waited outside for the arrival of the attorneys and their clients and witnesses.

The case was a civil suit without a jury, so members of the press were given the jury box to sit in. Placed on one side of the modern courtroom, these were the best seats in the house, comfortable leather chairs affording great views of a screen on the other side upon which exhibits were displayed. Jostling for a place in the box were print reporters, TV reporters, book writers, and documentary film makers from all over the United States and from many parts of the world.

To our left was the witness box, and beyond it the judge's bench, occupied by Judge John Jones. A George W. Bush appointee, Jones was a good-looking man in his fifties. In front of him sat the clerk of the court and the stenographer.

Directly in front of us was the lectern from which the lawyers asked their questions. Off to the right were the two legal teams, seated behind two large wooden tables. The defense was closer; the plaintiffs' lawyers were over by the far wall next to the screen.

Extending all the way to the back wall were two rows of uncomfortable wooden pews divided by an aisle. This was where the spectators sat, usually aligned behind whichever team they supported.

Opening arguments were made by the lead attorneys for each side. For the plaintiffs, Eric Rothschild claimed that intelligent design was not science, that it had been thrust into the classroom for religious reasons, and that its presentation at Dover High was a contravention of the First Amendment's Establishment Clause.

Patrick Gillen, the Michigan-based lead attorney for the defense, portrayed the case as being solely about free inquiry in education. The school board members had no religious agenda, their only purpose was that of "enhancing science education," and the only thing that had changed at Dover High was the presence of a new book in the library to which students were directed by a modest four-paragraph statement.

. . . .

The rules of the game in a civil suit are as follows. The plaintiffs—in this case, the eleven parents suing the school board—present their case first. Each of their witnesses undergoes a direct examination by one of their own lawyers, followed by a cross-examination by one of the lawyers from the other side, the defense. These are generally referred to as simply "direct" and "cross" and are sometimes followed by a "redirect" and a "recross" and occasionally a "re-redirect" and a "re-recross." When the plaintiffs finish presenting all the evidence they can extract from their witnesses (their case), the defense—in this case, the school board—then presents all its witnesses in identical fashion.

At the end of it, the judge looks at all the evidence and reaches a conclusion. It was interesting in a case about science to see how close the judicial method was to the scientific method: hypothesis (the plaintiff's case), antithesis (the defense case), and thesis (the judge's ruling.)

All the eleven plaintiffs were parents of children in the Dover school system. Seven of them were women; four were men.

Tammy Kitzmiller became the lead plaintiff because she had a daughter in the ninth-grade science class where intelligent design had been presented. She was a divorced mother of two and worked for a landscaping company as an office manager.

There were four teachers, one woman and three men. Kitzmiller's next-door neighbor and fellow plaintiff owned and ran a small electrical repair business with her husband. There was a legal assistant, an administrator at a youth advocacy program dealing with troubled kids, and a medical technologist.

And then there were Aralene "Barrie" Callahan and her husband, Fred.

One of the first plaintiffs to take the stand, and the one most qualified to talk about the early days of the school board, Barrie was a short, solid woman with a mop of densely curled blonde hair, from under which her blue eyes sometimes took on a glowering aspect. She reminded me of a female Harpo Marx who had finally gotten pissed off and wasn't prepared to play the clown anymore.

The wife of a local businessman, she wore colorful but well-tailored clothes and lived in a house up on a slight hill outside Dover. She had a degree in psychology, was a gourmet cook, and, as I would find out when I visited her, drank good wine. She and her husband, Fred, also a plaintiff, had three children. At the time of the trial, their youngest, Katie, seventeen, was in Dover High. Aralene, their oldest child (known as Arrie), had graduated from college and was about to go to Burkina Faso to teach English and theater. Their son, twenty-year-old Danny, was about to leave to spend a year in college in Australia.

Fred was a trim, good-looking man, a gentle man and a gentleman, but not someone you would ever push around. Always impeccably turned out, often in tweed jackets, his hair well cut, he was as concise, polite, and measured as Barrie was outspoken and incensed.

Barrie Callahan's direct examination was done by Steve Harvey, Rothschild's colleague from Pepper Hamilton.

A prematurely gray-haired man in possession of the best suits in the trial, Harvey was born in Pittsfield, Massachusetts, in 1960 into a large, staunchly Democratic family. His father was an engineer for General Electric, his mother a housewife. He went to a Catholic school and then on to public school in the seventh grade.

After graduating from the University of Massachusetts, he sold slacks in a department store, became a bank teller, and then got a job in New York with a private fundraising company, working largely for Catholic charities. After a four-year break from academia, he went to law school at Villanova, a Catholic university, and graduated in 1989.

He law-clerked for a judge in North Dakota—"It was cold"—and then got a job with the Justice Department in Washington, DC, in the Civil Division, defending the government against accusations of unconstitutional or unlawful actions. He was involved in defending the White House in the case of the Iran-Contra tapes.

Harvey began to feel like a Republican during the Clinton scandals, and when he moved to Philadelphia to join Pepper Hamilton and saw all the corruption in the city's Democratic institutions, he broke with family tradition and registered as a Republican.

He was a man full of contrasts. He was in his second marriage and had two young children. As a practicing Catholic he was against abortion but thought it a tricky situation legally. He was not opposed to contraception, thinking, in fact, that the Catholic church's attitude to it, particularly in Africa, was ridiculous. An ambitious man who wanted to do well in corporate law, he also did a lot of pro bono work on prisoner civil rights cases and now directed a legal clinic for the homeless.

He had not given intelligent design any thought at all until Eric Rothschild came to visit him in his office. He had worked on several cases with Eric and liked and admired him, so he immediately expressed interest. However, given the nature of the case, he thought he should do some research and think about his own philosophical relationship to the subject. At first he thought intelligent design was probably harmonious with his views, because basically all it was saying was that the wonder of nature suggested the existence of God.

"I think that's a reasonably good argument—but then when you say,

'Oh, no we're making this argument as a matter of *science*,' you know this is not just an appeal to intuition; they're saying they can prove it as *a scientific proposition* ... well, that's obviously a totally different thing, and that to me is what makes it such crap."

The most light-hearted of all the lawyers, he was also the most fearsome. His considerable personal charm and boyish smile disappeared entirely during cross-examination to be replaced by a cold, steely-eyed intensity that was absolutely chilling.

He and Eric complemented each other well. Eric was deceptively playful, even self-mocking at times. He had plenty of scalpels in his pocket but preferred to just humorously clink them together and produce them only when necessary. Harvey wore his polished knife in plain sight.

With Barrie Callahan, of course, none of this was evident. As Harvey led her through her testimony he was the charming Irish gentleman.

Callahan had served on the school board from 1993 to 2003. In a way, she had probably been, though inadvertently, a large part of what caused the debacle, because, behind the religious ideology that eventually consumed the board, many of its members initially stood for election or re-election on a thoroughly down-to-earth platform of fiscal conservatism and had done so in opposition to what they considered an over-ambitious building program at the school, of which Callahan was a big supporter.

By the late nineties, the school was in pretty bad shape. Classrooms had missing ceiling tiles, halls lacked floor tiles, some of the rooms were either too hot or too cold, the auditorium was not large enough to seat the entire student body, and the library was long past due for modernization.

In October 2001, investment advisers told the school board that in the wake of 9/11, this would be a good time to borrow money for what had been budgeted as an $18.7 million renovation.

Alan Bonsell, the owner of an auto repair shop and a key player on the fundamentalist side, objected to the project because it would raise property taxes. During later testimony he would say that the people of Dover liked to call Barrie's plan "a Taj Mahal version of the high school." According to him, the board was "talking about spending thirty, forty million dollars on a high school building project, which a lot of us in the community felt was ridiculous, being in the kind of tax revenue situation that we have in Dover."

In November 2001, Bonsell and a woman named Angie Yingling joined Sheila Harkins and Carol Brown, who were up for re-election to the school board on a platform of fiscal responsibility. Their campaign slogan was something along the lines of "Taxed to Death."

They were elected, joined Barrie Callahan on the board in December, rapidly scuttled the more expensive building plan, and, much to Barrie's irritation, started over. But not fast enough for the students. In May, a hundred and fifty of them walked out of class to protest the school's continuing dilapidation and its board's inability to do anything about it.

Although a district task force had already rejected the earlier plan as being inadequate, the board now voted eight to one (you can guess who the one was) to have a local architect draw up plans for an even lesser $9 million renovation. (The eventual renovation would cost over $20 million and end up being 30,000 square feet smaller than the original plan. According to Callahan, there is no new library, and the auditorium is still not large enough to contain all the kids at the same time.)

At first, Alan Bonsell was on the school board's curriculum committee, and at a board retreat early in his tenure, he spoke of wanting to have creationism taught at Dover High. But quite aside from this, the mood at the new board meetings was tense from the start. In the spring of 2002, two male board members resigned in frustration. This meant that the board could now appoint whomever they wanted to replace them.

Jane Cleaver, longtime owner of a five-and-dime in Dover, was appointed along with Bill Buckingham, the ex-cop and corrections officer who would go on to make the statement about Christ on the cross. Cleaver would support Buckingham and Bonsell in their attempt to get intelligent design into the science class.

Callahan stayed until her term ran out in 2003. She was not re-elected. Replacing her was a woman named Heather Geesey. Like Jane Cleaver, Heather would also support intelligent design. This meant that at the start of 2004, four board members were amenable to an idea that had been on Alan's mind since he first arrived.

There were no atheists among the school board members. Even school board member Jeff Brown who voted against intelligent design and became perhaps the most vocal opponent of it, was a Sunday School teacher.

After Callahan left the board, she remained involved in school matters and soon got into a public and protracted battle with the board over the nonpurchase of a biology textbook. The school's edition of *Biology* by Miller and Levine was several years old, and Callahan, whose daughter would be using it soon, felt it was time to buy an updated version. For some reason, no matter how hard she pushed, the curriculum committee never seemed to get around to making this purchase...

When Harvey had finished leading Callahan through the above history, Patrick Gillen, lead attorney for the defendants, stood up to cross-examine her. Patrick worked for the Thomas More Law Center, based in Ann Arbor, Michigan, which had agreed to represent the school board pro bono.

Run by Richard Thompson, a conservative Catholic and former public prosecutor, the center was originally financed by Domino's Pizza founder, Thomas Monaghan, also a Catholic. According to an article in the Sunday *Detroit News,* Monaghan gave an initial gift of $500,000 to get the law firm going and continued dumping cash into it until the middle of 2004. Since then, its annual budget of just over $2 million has been raised through fifty thousand individual donors, each paying $25 annual membership fees. The shortfall of over $800,000 is taken care of by other benefactors.

The Thomas More Law Center's motto is "The Sword and Shield for People of Faith," and its stated missions are "Defending the Religious Freedom of Christians," "Restoring Time Honored Family Values," and "Protecting the Sanctity of Human Life."

As a biblical literalist myself, I assume this must include such biblically mandated religious freedoms as the right to capture a woman in battle, shave her head, and lock her up for a month; time-honored family values that then allow you to rape her into matrimony; and, finally, having done all this, protect the sanctity of life by refusing her the right to an abortion. (For all but the last, see the book of Deuteronomy.)

Expanding on the religious freedom aspect, the website says:

> We live in a culture increasingly hostile to Christians and their faith. America has become a nation where public school students are prohibited from praying, acknowledging their dependence on God, and forming religious clubs . . . Our Founding Fathers fought for a nation built on a foundation of religion and morality. Our lawyers are committed to restoring and preserving that foundation.

Before the Dover case, Patrick Gillen had been involved in an antiabortion case in San Diego and a protracted lawsuit in Michigan that was trying to prevent the local school district from providing insurance benefits to same-sex partners. This lawsuit was, as the Thomas More Law Center proudly proclaimed,

> just one of several related efforts by the Thomas More Law Center to defend traditional marriage and block homosexual activists from acquiring the benefits of marriage that would result in the de-facto legalization of same-sex marriage. The Center has worked to stop transsexual marriage in Kansas, homosexual adoption in Nebraska, and is involved in major legal battles over the rights of students and faith based organizations to reject homosexual demands for recognition of their lifestyles and unions.

As someone who abhors homophobia, this information did not stimulate my affection for Patrick Gillen. However, one of my chief defects—or better qualities, I'm not sure which—is that I find it almost impossible to maintain animosity toward people with whom I violently disagree once I get to know them. Usually, in fact, I only have to look into an enemy's face to find something I can sympathize with. With Gillen, my sympathy was ignited by the contrasts. A tall man, probably in his late thirties, whose long head was topped with thinning hair, he had excellent teeth, revealed frequently in a blazing grin; but from the middle of his nose up—and all this at the same time—he wore an expression of extreme anxiety, his brows furrowed, his eyes filled with concern.

Later in the trial he would come under tremendous pressure (often with little or no support from his co-counsel or legal assistants, who came and went with inexplicable frequency), but Patrick was never rude or even impatient. Everyone liked Patrick, myself included. He was so unfailingly polite and sweet that I was not alone in wondering whether he really belonged in an environment as harsh and intolerant as the Thomas More Law Center.

In his brief cross-examination of Barrie Callahan, Patrick established that issues raised at board retreats were not voted on and that Alan Bonsell had not, as far as she knew, tried to implement his desire to have creationism taught in 2003. He got Callahan to concede that issues of "fiscal concerns" might have played a part in the nonpurchase of textbooks.

When he began to question her about how she came to believe intelligent design was not science, and whether or not she shared anything she had learned about it with other board members, he got more than he bargained for.

At one board meeting she had distributed copies of a *National Geographic* article containing a definition of "theory." At another she had brought copies of an article from the *New York Times* Sunday magazine.

"I think 'The Genesis Project' was the name of that article. And it talked about a lot of the scientific discoveries behind origins of life. I mean I can go a little bit into that, if you would like ... If you'd like me to, I can."

"No, that's all right; that's fine," said Gillen.

Bill and Alan

Although other school board members would support them, the driving forces of the intelligent design/creationism movement in Dover were Alan Bonsell and Bill Buckingham.

Bonsell was a good-looking, sandy-haired, gum-chewing man somewhere in his late thirties or early forties. He wore casual clothes that looked expensive and had the relaxed, entitled, slightly contemptuous air of a politician or an athlete. He sat throughout the trial, which he visited most often in the afternoons, lounging on the uncomfortable pews, arms outstretched behind him, head back, often grinning.

He reminded me of President Bush in some ways. His faith seemed to have given him a confidence unwarranted by the facts. He had a degree in business management from York College, but his auto repair shop displayed few signs of either management or business. In a dismal section of town, it was hard to imagine that it generated any significant income. Alan was, however, relatively wealthy, as was his father, Donald Bonsell.

The source of their wealth, I was told, was real estate, and when I checked the county records, I found the Bonsells did indeed own many properties. Donald lived in a large old house, Alan in a large modern house in a development at the entrance to the local golf club. They both had the reputation of being bullies—perhaps Donald, the father, who at one time was the town supervisor, more so than Alan. But even Alan, according to one of the plaintiffs, became so angry on one occasion that he threatened physical violence.

According to all who knew him he did not drink, and was, depending on who you spoke to, either deeply religious or a sanctimonious, hypocritical prig. He was also a man who would, according to the highest authority in the court, lie under oath.

The church he attended, the Church of the Open Door, declaims on its website that "The Bible is the very Word of God and is the only infallible rule of faith and practice." "Jesus was born of the virgin Mary and gave Himself as a ransom for the remission of the sins of man." "All persons are lost apart from regeneration and are condemned to eternal punishment, while the saved even now possess eternal life." And last but not least, "The second coming of Christ is personal, premillenial, and imminent."

While reading this, I remembered something another school board member, Jeff Brown, had told me when I interviewed him. I don't know whether he was talking about Bonsell specifically, but I suspect he was.

"These people are convinced," he said, "that evolution—and they have said this to me—they're convinced that evolution is a hoax, that science has evidence debunking it, but they suppress it." What they believe instead is "that we've got to return to our biblical roots because the end of times is coming. And when you take that viewpoint, then I can understand why it would be very easy to believe that what you learn in ninth-grade biology class is really unimportant because in just a year or two you're going to be either cast into hell or whisked off to heaven, where it isn't gonna matter what you studied in school anyway. The important thing is to get those souls saved."

Alan unashamedly believed in a literal interpretation of Genesis. With a smirk of disdain, he said that he found evolution "absurd," dismissing the whole matter with the kind of irritated shrug one might offer to someone suggesting that JFK was bumped off by aliens.

If Alan was the new, entitled, middle-class, Bush-empowered face of fundamentalism, Bill Buckingham was not. He was an unapologetic, old-time, tell-it-like-it-is, hardcore American fundamentalist and an outspoken "nuke Iraq" patriot.

Born in 1946, he looked older than his years. When he came into court later in the trial, he wore a light brown camel hair sport coat, on the lapel of which was a pin consisting of a cross wrapped in an American flag. For reasons that will become apparent, he walked using a cane. He was, at least before and after the trial, though not so much during it, a bumptious and aggressive gadfly who seemed to make outlandish statements almost for the fun of it. He knew nothing about evolution or intelligent design, and when asked to give a definition of the latter he gave a pretty fair definition of the former.

He had grown up in the area and graduated from William Penn Senior High in York on June 2, 1964. By June 4, he was in boot camp in Parris Island, South Carolina, training to be a Marine. In 1969, he returned home and became a cop in York.

This was a brave or reckless career choice. At the time York was in the throes of ferocious race riots, the repercussions of which were still felt in the district when I was down there. In 2001, over thirty years after the riots, the mayor of York, Charles Robertson, a former police officer, was arrested and led out of City Hall in handcuffs. He was accused of providing ammunition during the riot to white teenaged gang members and urging them to go out and kill as many black people as they could. They killed a black preacher's daughter. In 2005, both Robertson and the teenagers went on trial. The latter were found guilty and sent to prison. Robertson was acquitted by an all-white jury just before *Kitzmiller v. Dover* began.

Buckingham became a detective in the narcotics division. York was a rough town and being a cop in it was a rough life. He was shot at by drug dealers and peppered once with pellets. In 1977, it got to be too much. For the sake of his family, he decided to leave the police force, and in 1977 he went to work at York County Prison, where he became a supervisor.

In 1982, he was working a short-handed shift in the visiting area when he saw a prisoner slip another a piece of paper. When he demanded to see it, the man resisted.

"It was a set-up," Buckingham told me. "The deal was—back then I was a weightlifter, I knew karate, and I could have handled myself very well—they were hoping I would hurt somebody and they could file a lawsuit against the county. But what happened is, I didn't do that. We just got into a wrestling match . . . The prison is made out of steel and concrete, and I was bounced off the corner where the steel met the concrete, and it hit me in the low back."

When he finally got help, he was so angry he actually picked up one of the guys, who weighed over two hundred pounds, and carried him into a holding cell. It was not until he returned to his office and sat down that his back began to hurt. He was, in fact, so badly injured that his working life was essentially over before he reached the age of forty.

Over the next few years, he had six operations on his back, during the last of which "I was literally cut in half. When I woke up in the recovery room and they said 'rate your pain from one to ten,' I begged them to kill me . . . I told my wife, 'If I ever talk about back surgery again, you remind me of this, 'cause I'll die first.'"

At the time of his final operation, he was a eucharistic minister at a local Catholic church. "I helped to give out communion and so forth," but no one from the church ever came to see how he was doing. One day "a pastor at the church where I go now stopped where I had coffee once in a while and we talked . . . and he invited me up to his church, 'cause I wasn't going anywhere. And I went up and that was the first time I felt like I'd been to church in a long time, and I stayed."

He was born again in 1993. When I asked him how it changed him, he said he "cared more for people" and spoke of how he and his wife sang Southern gospel hymns and led prayer groups in local hospitals and nursing homes.

By 1998, he was in so much pain that he was prescribed OxyContin.

Within a year or two he was an addict.

Perhaps in part because of his addiction, or the painful consequences of trying to escape it, the looming mist of his fundamentalism was, in early June of 2004, about to congeal into something angrier and harder—something hard enough, in fact, to hit the blades of the proverbial fan and send an audible ping of alarm out into the wider world.

Life on Board

IN MY MANY trips into the interior—of the United States—I have been pleasantly surprised to discover that no matter where you go, you will almost always find one or two unapologetic, intellectually curious eccentrics.

Jeff and Carol (often known as Casey) Brown would certainly qualify.

Jeff, a chain-smoking mustachioed electrician who wore thick glasses and a floppy tweed cap, reminded me of a jazz drummer I once knew who lived a strenuous yet enjoyable life but had experienced so many unusual states of consciousness and read so many odd metaphysical books that, although he was fascinating to listen to, he was not altogether capable of what customarily passes for conversation. Jeff was a Sunday School teacher and, I'm willing to bet, one of the most interesting you could ever encounter. He required only the lightest of nudges to philosophize on any subject you cared to choose.

Carol was a striking woman with reddish hair, who held herself well. Shyer than her husband, she read the *New York Times* through thick lenses and conveyed an air of edgy detachment, as if her long inhabitation in an area not noted for its admiration of the brain had made her feel both superior and slightly anxious. She spoke in a refined manner occasionally contradicted by a hoarse, knowing laugh. She had attended several colleges, had a degree in secondary education, and had done some graduate work "pursuant to archeology." For a few years she was a reporter for the *York Dispatch* and the *York Sunday News.*

When Jeff, who attended Dover High, was asked in his deposition about his education, he replied, "Beginning with high school?" "Yes." "It also ends with high school, just for the record ... In those days, it was grade seven through twelve—it was all called high school. There was no junior high ... You went to elementary from first to sixth and high school from seventh through twelfth. And at that point, I terminated my education."

The Browns lived in a rambling old house cluttered with books and other mementos of a life of intellectual curiosity. As a hobby, Jeff created comic books: words and pictures. The one he showed me when I visited was well executed and brought to mind the more rugged adventure stories of Arthur Conan Doyle, with the addition of buxom women straining at their blouses as they tramped through fantastic and dangerous terrain.

Both he and Carol had children from previous marriages who attended school in Dover. The Browns had not met until later in life, an event that Jeff, in spite of an inherently laconic drawl, recounted in enraptured detail like a man describing the discovery of a refreshing oasis while trekking (without a straining-at-the-blouse companion) across a desert. They stayed up all night talking. Having spent more time with him than with her, I imagined he did most of the talking until I looked at his deposition and saw that when Patrick Gillen asked him how long he and Carol had met with the plaintiffs' lawyers the night before, he replied, "About two hours. Since you have already deposed my wife, you can guess why it took so long."

Both were on the school board for many years—she for nine, he for five—until both resigned during a wildly fractious board meeting in the fall of 2004. Carol, though not a plaintiff herself, was the plaintiffs' best witness to the inner workings of the board as it slowly and painfully self-destructed.

Her direct examination was done by Eric Rothschild, the plaintiffs' lead attorney.

In his late thirties, Rothschild was a short man with a balding head shaved close. He had a deceptively cherubic face, which, on closer inspection, was seen to be a dark face too, with the air of someone keeping a secret. One might imagine that as a geeky child he had encountered some bullying and had decided at a certain point that he was not going to let it continue into adulthood.

He was a second-generation American and a practicing Jew, who took his kids to temple. His parents were German, and he had been raised for the first few years of his life in Bonn, where his father worked for the Central Intelligence Agency. When Eric was about to go into the fifth grade, his father was posted to CIA headquarters in Langley, Virginia. Rothschild went to public school in Maryland.

From high school, he went to Duke University. At 19, he met his wife, Jill, now a pediatrician. When she came up to Philadelphia to attend medical school, he followed. He waited tables but also did some work as a legal assistant. After a while, he decided to go to law school, applied to the University of Pennsylvania, and got in.

In 1994, he joined Pepper Hamilton and had, with one short break, been there ever since. Most of his work was in product liability and commercial litigation. His first case at Pepper Hamilton was perhaps the most relevant to Dover: the defense of the Three Mile Island nuclear plant from a long-running lawsuit brought by people claiming to have been damaged by a meltdown and subsequent leak in the seventies.

When I went to visit him at his home in a suburb of Philadelphia, his wife remarked that she and Eric were both liberal Democrats and this was not the kind of case either would naturally have sympathy for. "We got a lot of jokes about this from our friends," laughed Jill. "I remember him trying to convince me . . ."

I already knew that the Three Mile Island nuclear plant lay only thirty miles from Dover and had become convinced the wind was blowing Dover's way on the day of the radiation leak. This, I believed, explained the weird behavior in the now grown-up children sitting on the school board. Sadly, Rothschild disabused me of this. The leak had been small. They won the case.

Carol Brown started out by telling Rothschild who was on the board when intelligent design tore it apart: board president Alan Bonsell, Carol herself, her husband, Jeff, Bill Buckingham, Sheila Harkins, Jane Cleaver, Heather Geesey, Angie Yingling, and Noel Wenrich.

"Do you consider any of these people your friends?" asked Rothschild.

"I did, sir," she replied in a tone of bitter regret.

"All of them?"

"Yes, sir."

Her first inkling that there might be a problem came at an evening board retreat in January 2002, three weeks after Alan Bonsell had been elected for the first time. "He was very concerned with the state of morality," she said, "and he expressed a desire to look into bringing prayer and faith back into the schools . . . He mentioned bible, and he mentioned creationism."

Richard Nilsen, the Dover School superintendent, had taken notes of all the board members' concerns, which were now put on screen in the court. Under Alan's name was written "Creationism, Prayer, Need administration to work as a team, Curriculum, Uniforms."

In March 2003, there was another board retreat, at which Richard Nilsen again took notes. This time he was accompanied by Michael Baksa, the new assistant superintendent. By now Alan Bonsell was vice president of the board and chairperson of the curriculum committee. Once again, he mentioned creationism, but now he seemed equally concerned that the students were not learning enough about the Founding Fathers and early American history.

Rothschild asked Brown whether until this point the Dover schools had been teaching about the Founding Fathers and the early colonial period. She replied that yes, they had, but "there was not the emphasis that I believe Mr. Bonsell wanted to see."

"And what specifically was that emphasis?"

"His emphasis was more on making our students aware of the importance of faith in the early history and founding of our country, sir."

Around this time, the science department, aided and abetted by Barrie Callahan, started to press for new science textbooks, among them the previously mentioned *Biology*.

"We had an extremely tight budget," Brown explained. "Not that we always didn't, but it was very much so that year. And even though it was the cycle time for science books, we put off purchase for one year."

In the 2003/2004 school year, the teachers renewed their request. By this time, Alan Bonsell had become president and appointed Bill Buckingham to chair the curriculum committee. Buckingham soon came to hate the biology book, the co-author of which, Ken Miller, would be the first expert witness in the trial.

Miller Grinds It Up

Witold Walczak, the legal director of the Pennsylvania ACLU, was one of the sharpest-elbowed lawyers in the case.

Unrelentingly combative, he had the weary but determined air of a man who spent his life, for little pay, passionately defending the Constitution, but knew that the only questions most people would ask him would relate to the ACLU's defense of NAMBLA (the North American Man Boy Love Association) and the Ku Klux Klan, the latter of which he had personally defended on several occasions.

Known as Vic, he was, like Rothschild, a second-generation American. His father was Catholic; his mother was Jewish; both were Polish. His grandmother died at Treblinka. His grandfather, a lawyer and violinist, was at the same camp. Rather than kill him, the Nazis surgically removed the tendons from his bowing arm.

The surviving Walczaks left Poland in the fifties, sought asylum in Sweden, where Vic was born in 1961, then eventually moved to the United States, settling in Scotch Plains, New Jersey—or, as it was affectionately known, Crotch Pains.

After high school, he went to Colgate, a college in upstate New York, and while there took a class that changed his life—"Leaders in non-violence"—about the lives of Leo Tolstoy, Mahatma Gandhi, and Martin Luther King.

Vic became a philosophy major with a concentration in ethics.

In his junior year he took time off and went to Washington, DC, where he worked as an investigator for the public defender's office before talking his way into a job with a social service agency that worked with juvenile delinquents in some of the worst areas of the city.

"I remember one night they had some outdoor event up on 14th and Fairmont in northwest DC, which was the center of the drug trade, and I'd been working with a kid there who'd been having a lot of problems . . . I'd been there for four or five hours, playing basketball, sitting on the steps, talking to people. I looked around and realized I'm the only white person probably for three square miles—and it hadn't entered my consciousness. It was really a wonderful feeling."

After graduating from Colgate, he went to Poland to stay with a large family in Gdansk, the center of the Solidarity movement and home to Lech Walesa, its leader, a shipyard electrician. The day he arrived, he had five oranges in his bag. Within minutes they disappeared. It was soon discovered that one of the kids, a twelve-year-old boy, had stolen them and eaten them all. His mother was embarrassed but explained that the child had eaten an orange only once before in his life.

The father came back from the shipyard where he worked, and after dinner he and Vic went for a walk. They passed a food store filled with chocolate, fresh cuts of meat, and fresh fruit, including oranges. Vic was surprised: hadn't he just been told the son had eaten only one orange in his life? The man explained that you could shop there only if you belonged to the Communist Party.

Walczak was outraged and said so. The man put his hand over Vic's mouth and told him to be quiet: you could not know who was listening. "And this was the start of understanding the paranoia everybody in a communist or totalitarian country has about saying something that displeases the government."

Later in his visit, he attended a Solidarity demonstration, during which he saw men being beaten. He took photographs and then had to run away quickly to escape the police, who wanted to take his camera. Coming after his experiences of poverty in Washington, the summer in Poland defined him: he wanted to defend and protect people's civil rights. He returned to the States and went to law school at Boston College.

Although he did not enjoy law school, he graduated with honors. His then girlfriend—now wife—was off to medical school in Washington, DC, so he followed her there and began working for a group that dealt with prisoners' rights. He filed so many class action lawsuits that prison wardens began making threatening calls to his boss. Walczak, they claimed, was being "irresponsible." One of his cases led to twelve lifers, among them murderers and rapists, getting out on parole. Before I could say anything, Vic smiled a little defiantly and said, "And they're all doing well."

After five years his wife, like Rothschild's, became a pediatrician. When she was offered a job at Children's Hospital in Pittsburgh, they moved there, and Walczak started looking for a job. One day he had an interview with a well-respected local lawyer, a woman with her own firm. After a while, she said, "You know, you don't strike me as law firm material, but I'm on the board of the local ACLU, and we had a meeting last night and created a one-year associate director position . . ."

He instantly applied, but, not daring to hope he might get the job, he kept applying elsewhere. By the time he was accepted by the ACLU, he had six job offers. "I took the lowest-paying job—and I've never looked back."

His office was in stark contrast to the corporate tower that Rothschild and Harvey inhabited. It was a rickety two-story house in a mixed student and blue-collar neighborhood.

In the months leading up to *Kitzmiller*, Walczak was involved in defending the civil rights of antiabortion protesters forbidden from displaying graphic pictures of fetuses on a highway overpass; an adult theater being shut down by zoning laws; a woman charged with disorderly conduct for holding up a sign at a country music concert saying "Show us your—" followed by a picture of breasts; a prisoner and a group of taxpayers alleging that government agencies were giving money to a faith-based work-release program that used the money to proselytize prisoners; Amish drivers who had religious objections to displaying orange triangles on their buggies; a girl who refused to stand during the Pledge of Allegiance; gay and lesbian employees of the University of Pittsburgh denied domestic partner health benefits; a woman prevented from seeing her son in the hospital after he was shot by the police; a boy kicked out of school for writing a violent rap song; and a woman who put a 32-foot-long antihunting painting on her front lawn depicting Bambi getting shot.

Walczak was a true believer, and what he truly believed in was the Constitution. If you didn't defend the civil rights of even people you disagreed with, the Constitution meant nothing. While Eric Rothschild and his Pepper Hamilton colleague, Steve Harvey, left court after brief, tight-lipped statements, Walczak would often hang out with the reporters after the trial, standing on the street, exhausted, back to a wall, to share his scrappy, sardonic anger at the dishonesty of the opposition. He was the most hated of the plaintiffs' lawyers. One school board member in Dover was heard to say, "I fear the ACLU more than I fear Al-Qaeda."

Walczak was used to this animosity and had in fact developed a strategy—which rarely worked—for protecting his family from it.

"I don't advertise what I do," he told me when I went to visit him in Pittsburgh. "I won't say, 'I'm the legal director for the ACLU of Pennsylvania . . . we're the most important organization in this country that ensures we have real justice and fairness for everybody.' No, I'll say: 'I'm a lawyer.' 'Oh, who do you work for?' 'For a public interest organization.' 'What kind of law do you do?' 'I do civil rights law.' 'Oh, that's interesting—do you work with a firm?' You know, when you get to the eighteenth question: 'ALL RIGHT, I WORK FOR THE ACLU!!!'"

. . . .

Kitzmiller v. Dover had two distinct legal themes. One was the story of neighbors in conflict in Dover, the other the clash of opposing scientific and philosophical views on the origin and development of life.

In the first of these, the question was this: did the Dover advocates of intelligent design in fact have religious motives, and even if they did not, did their actions result in a religious theory being injected into a public school science class? (The words "purpose" and "effect" were key to this aspect of the case, much of which would eventually hinge on questions of human honesty—or lack of it.)

The second theme, the scientific and philosophical one, swirled around the question of whether or not intelligent design was in fact a religious theory, "creationism in a lab coat," and whether, even putting that aside, it was good science.

Was it, in fact, science at all?

The plaintiffs' first expert witness, Ken Miller, co-author of the textbook *Biology* and a professor of biology at Brown University, was here to address that question, with help from Walczak.

As the plaintiffs had to present their case first, their scientific witnesses had to explain and describe intelligent design before they could criticize it. It was Ken Miller's job, therefore, to throw a rope over a branch of the tree of knowledge, diligently hang the piñata of intelligent design from it, and than proceed to whack at it for the next day and a half.

The foremost scientific proponent of intelligent design was a man named Michael Behe. A biochemist and tenured professor of biology at Lehigh University in Bethlehem, Pennsylvania, where he had taught for twenty-three years, he would later appear for the defense as their leading expert.

Behe did not invent the term "intelligent design," but he did coin the expression "irreducible complexity," and he offered the best definitions of these interlinked ideas.

In *The Origin of Species*, Darwin wrote, "If it could be demonstrated that any complex organ existed which could not possibly have been formed by numerous, successive, slight modifications, my theory (of evolution) would absolutely break down."

Irreducible complexity took up that challenge. "By irreducibly complex," Behe writes in his book *Darwin's Black Box*, "I mean a single system which is necessarily composed of several well-matched, interacting parts that contribute to the basic function, wherein the removal of any one of the parts causes the system to effectively cease functioning.

"An irreducibly complex system cannot be produced directly ... by slight successive modifications of a precursor system, because any precursor to an irreducibly complex system that is missing a part is, by definition, nonfunctional. An irreducibly complex biological system, if there is such a thing, would be a powerful challenge to Darwinian evolution. Since natural selection can only choose systems that are already working, then if a biological system cannot be produced gradually, it would have to arise as an integrated unit, in one fell swoop, for natural selection to have anything to act on."

Another way this was explained was through the analogy of the mousetrap. If you removed any part of a mousetrap, it ceased to work. Furthermore, none of its parts, so the I.D. proponents claim, would have reason to come into existence because none would have had a previous function of its own. Ergo, an intelligent designer was required to create the thing "in one fell swoop" in its present form for its present function.

Miller began his testimony by defining "science." "Science" comes from the Latin word *scientia,* meaning simply "knowledge." But in the context of the trial, what was meant by "science" was "natural science," sciences like chemistry, biology, physics, and astronomy. He used a publication from the National Academy of Sciences to describe the modern practice of science:

> In science, explanations are restricted to those that can be inferred from confirmable data, the results obtained through observation and experiments that can be substantiated by other scientists. Anything that can be observed or measured is amenable to scientific investigation. Explanations that cannot be based on empirical evidence are not part of science.

Miller suggested that these rules had, in a way, always applied to science, but certainly in a formal sense for the past two hundred years, and they applied to all scientists everywhere in the world.

"I think science might be the closest thing we have on this planet to a universal culture," he said.

In trying to explain why confining science to these rules was important, Miller, a Boston Red Sox fan, said he might state that "the reason the Boston Red Sox were able to come back from three games down against the New York Yankees was because God was tired of George Steinbrenner and wanted to see the Red Sox win . . . And you know what, it might be true, but it certainly is not science, it's not scientific, and it's certainly not something we can test."

In other words, science, by self-regulation, admits it is not capable of dealing with issues of meaning or purpose. That is the job of philosophy or theology. All science does is ask questions about the natural world, propose hypotheses, and then seek to test them. If a hypothesis fails, it is discarded. When evidence supports the hypothesis, it is still not considered proven, merely supported.

If these are the rules of science—and most people acknowledge they are—it is strange it has such a reputation for arrogance, particularly when compared with religion, from whence this criticism often comes.

Miller now went on to outline evolution's three main hypotheses:

That the natural history of the planet is characterized by a process of change over time. That living things, if traced back far enough, show common ancestors that gave rise to the many forms of life today. And that to a large extent the process that drove change through time is natural selection.

Natural selection was Darwin's most significant contribution to the theory of evolution (parts of which existed before he was born), and, paradoxically, seems the most obvious. As Darwin's friend Thomas Henry Huxley said when Darwin explained it, "How extremely stupid not to have thought of that!"

The process is often portrayed as being wholly brutal, but, as Miller pointed out, part of the effort to survive involves the struggle to find mates, to reproduce, and to raise those offspring. These endeavors are anything but brutal, often requiring a great deal of collaboration and cooperation.

Contrary to what Behe said, Miller claimed that since Darwin's time, new developments in science (modern genetics and molecular biology in particular) have consistently confirmed his theory.

One example was the discovery that the great apes have forty-eight chromosomes, while humans have only forty-six. Chromosomes are far too valuable simply to be thrown away. In fact, to be lacking one would probably be fatal. "Therefore, evolution makes a testable prediction, and that is that somewhere in the human genome we've got to be able to find a human chromosome that actually shows the point at which two of these common ancestors [ancestral chromosomes] were pasted together."

As chromosomes have little genetic markers at their middles and ends, these should make the identification of the spliced ancestor gene even more definitive. And "lo and behold," to quote the professor, a recent peer-reviewed article found that "the answer is in Chromosome Number 2 . . . All of the marks of the fusion of those chromosomes predicted by common descent and evolution are present."

To give a perspective on the value of evolution in modern life, Miller read from another publication of the National Academy of Sciences.

> The concept of biological evolution is one of the most important ideas ever generated . . . The evolution of all the organisms that live on earth today from ancestors that lived in the past is at the core of genetics, biochemistry, neurobiology, physiology, ecology, and other biological disciplines. It helps to explain the emergence of new infectious diseases, the development of antibiotic resistance in bacteria, the agricultural relationships among wild and domestic plants and animals, the composition of the earth's atmosphere, the molecular machinery of the cell, the similarities between human beings and other primates, and countless other features of the biological and physical world. As the great geneticist and evolutionist Theodosius Dobzhansky wrote in 1973, "Nothing in biology makes sense except in light of evolution."

With this in mind, Miller ripped into *Of Pandas and People*, calling the book "inaccurate" and in many respects "downright false."

Much of intelligent design theory rests on three examples of irreducible complexity: the bacterial flagellum; the mechanism by which blood clots; and the immune system.

Miller showed that in fact none of these were irreducibly complex. There were examples of all three functioning with a "missing" element, sometimes, as in the case of the blood clotting system of the puffer fish, with several missing or, presumably, having never developed. If we don't yet understand exactly how these organisms and systems evolved into more complex forms, science is working on it and proposing many viable hypotheses. *Of Pandas and People* seemed unaware of any of them.

The book (and Behe), while making numerous arguments *against* evolution, never produced any positive evidence *for* design. Furthermore, even if evolution was proved to be wrong, this did not automatically make intelligent design right.

"The logic of picking out intelligent design, which is inherently untestable, and saying that any evidence against evolution is evidence *for* intelligent design employs a logical fallacy that I think most scientists reject."

He went on to say that as a practicing Catholic with two daughters raised in the church, he had no problem believing in both evolution and God.

The final aspect of his direct examination was to analyze the purpose and effect of the statement the school board now required ninth-grade science students to hear before they began studying evolution.

"The School District argues," said Walczak, that "you know, it takes a minute to read this statement ... What's the big deal? What's the harm in reading this to Dover School District students?"

"That's a very interesting point," said Miller, "and if they raised the issue, one might well turn around and say, well then why read it in the first place if it is of so little consequence? Then why have you insisted on doing this and why are you in court today? The only thing I can infer from turning that question around is that the Dover School Board must think this is enormously important to compose this, to instruct administrators to read it, to be willing to fight all the way to the court. They must think that this performs a very important function ...

"Do I think this is important? You bet ... First of all, it falsely undermines the scientific status of evolutionary theory and gives students a false understanding of what 'theory' actually means.

"The second thing is, it is really the first attempt or the first movement to try to drive a wedge between students and the practice of science, because what this really tells students is 'You know ... you can't trust the scientific process. You can't trust scientists. They're pushing this theory. And there are gaps in the theory ... You really can't believe them. You should be enormously skeptical.' What that tells students basically is, science is not to be relied upon and certainly not the kind of profession that you might like to go into.

"And thirdly ... by holding [intelligent design] up as an alternative to evolution, students will get the message in a flash. And the message is: 'Over here, kids, you got your God consistent theory, your theistic theory, your Bible friendly theory, and over on the other side, you got your atheist theory, which is evolution ...'

"What this does is to provide religious conflict into every science classroom in Dover High School. And I think that kind of religious conflict is very dangerous."

• • • •

Miller's cross-examination was conducted by Robert Muise, another lawyer from the Thomas More Law Center.

Muise was a tall, sturdy man, quietly resolute, with a faint Boston accent. Always willing to talk, as unfailingly polite as his co-counsel, Patrick Gillen, he seemed like a tough guy, but worn down, becoming a victim. Perhaps this had nothing to do with politics and religion; he and Patrick, though both only in their late thirties or early forties, had seventeen kids between them. One had eight, the other nine.

Like Gillen, Muise had been involved in several cases trying to stop homosexual marriage. He represented an Ann Arbor Christian girl whose school prevented her from expressing her antihomosexual views during Diversity Week, and he had appeared with her on the TV show *The O'Reilly Factor*. Like the ACLU lawyer Vic Walczak he was a staunch defender of displays of graphic pictures of aborted fetuses, including, around the time of the trial, one on a banner towed behind a plane flying over Honolulu. The antiabortion activists were being denied, he claimed, their First Amendment right to "stir people to anger."

He was not so keen on the First Amendment's free speech clause when representing a teacher and a student at Washburn University who wanted an on-campus statue of a Catholic bishop, entitled "Holier Than Thou," removed from sight. According to Muise, it was "quite clear" that the bishop's hat looked like a phallus (and it did, if you studied it a moment—a large, fat phallus, in fact). The statue, he claimed—and I'm sure he was right—demonstrated a "hostility and disapproval toward Catholic beliefs and practices." Perhaps it was a hostile disapproval of Catholic priests molesting small boys, and the church's apparent "belief" that it was okay to protect these men when they got caught. For this case, Muise happily put aside the free speech rights he'd fought so hard for on behalf of antiabortionists and now invoked another aspect of the First Amendment:

"Just because it is labeled artwork doesn't mean it is not a violation of the Establishment Clause."

Of course, in cases to do with *reverent* displays of Christian paraphernalia on government property, the Thomas More Law Center once again reverted to the opposite view, and suddenly free speech was back in vogue.

If the defense thesis was that there was no religious motivation for teaching intelligent design in science class, that it was merely another scientific theory, one did have to ask oneself what these Catholic activists were doing in court. Didn't their very presence here reveal the lie behind the whole debacle?

Among the Thomas More Law Center lawyers, Muise was rumored to be the most enthusiastic about intelligent design; that in fact it was he who had been adamantly searching around for a case like *Kitzmiller*. He was certainly the one who in May of 2000 went down to Charleston, West Virginia, to try and get its school board to buy *Of Pandas and People*. He told the board they would probably get sued if they used it, but if they were, the Thomas More Law Center would defend them at no cost. The school board declined—as did school districts in Michigan, Minnesota, and "a handful" of other states, according to Thompson.

Thomas More finally got lucky in Dover, and here was Muise about to start mixing it up.

• • • •

Muise scored some of the best points for his side in his cross-examination of Miller and revealed what I regard as an essential weakness in arguments made by moderate believers of his kind. Now, however, for the balance of the day, Muise revealed more about himself and the people he represented.

Part of the problem was that Miller, a lifelong teacher, could take almost any question, divide it into segments, and give an erudite lecture on each one of them. But it was more than this. Almost in spite of himself, Muise kept revealing an antipathy to the very concept of science. Believing in certain absolutes, he seemed to regard science as sly and evasive, and if you felt this way, Miller would certainly appear to be its slyest and most evasive advocate.

The following exchange was fairly typical.

Muise: "Sir, would you agree that science involves a weighing of one explanation against another with respect to how well they fit the facts of experiments and observations?"

Miller: "I would agree that science involves the weighing of one *natural* explanation against another with respect to how well they fit the results from observation and experiment."

Muise: "Would you agree that all science consists of looking at the evidence and then drawing inferences from it?"

Miller: "I think that *part* of science is looking at the evidence and drawing inferences, but I hesitate to agree completely with your question because I certainly think that drawing just *any* inference from data is not necessarily scientific."

Muise made Miller define the difference between the word "theory" as it was popularly used—a hunch—and the scientific definition, which, Miller conceded, was expressed pretty well in the school board's statement:

A theory is defined as a well-tested explanation that unifies a broad range of observations.

But Muise could not seem to grasp this. "And as a theory, the theory of evolution is not a fact?"

No, Miller told him, it was a theory. It was made up of *lots* of facts that added up to a theory. In this sense, a theory was much *more* powerful and persuasive than any mere fact.

"Would you agree that Darwin's theory of evolution is not an absolute truth?"

Miller responded that no theory is ever regarded as absolute truth and this included atomic theory, the germ theory of disease, and the theory of friction. All were conditional.

Paraphrasing the school board's statement in the form of a question, Muise asked: "Because Darwin's theory is a theory, it continues to be tested as new evidence is discovered?"

Yes, replied Miller—as is so with all scientific theories.

And wasn't it true that "all of science was filled with gaps"?

"If you define an unanswered question as a 'gap,'" Miller said, with just a hint of disdainful impatience, "then it is certainly true that science is filled with unanswered questions."

Although this line of questioning was yielding nothing, Muise could not let go and asked several more: "Should we regard Darwin's theory as tentative?" "Darwin's theory of evolution is incomplete and unfinished; isn't that correct?" "So the origin of life is an unsolved scientific problem, correct?"

The questions he was asking—or, more accurately, the *accusations* he was making—instead of causing Miller shame were instantly transformed into compliments. Muise kept saying, "So you admit it! Science does not know everything!" and Miller kept replying, "Yes! Isn't that wonderful?"

So it went until the day was over.

It was unsettling. Muise seemed to equate Darwin and evolution with God and religion. The bible was the inerrant word of an infallible God. Unless Darwin was the same, he had no validity. For his theory to be acceptable, it could not merely be "a well-tested explanation that unified a broad range of observations" but had to be a *perfect* explanation unifying *all* things.

In the Scopes trial, William Jennings Bryan used an old chestnut to explain his scientific viewpoint: "I am more interested in the Rock of Ages than the ages of rocks." Muise was far better than this. He had learned whatever he could about the theory of evolution and the modern practice of science, and he brought forth every solid argument against them that he could. But, like Bryan, he had also, figuratively speaking, brought his bible, and what he simply could not grasp was that unlike his book, to which he was permanently glued as a matter of faith, scientists would toss *The Origin of Species* into the trash in an unsentimental second if a better theory came along—and in so doing would not be committing apostasy but upholding their most vital sacraments.

. . . .

Having spent the preceding day attacking evolution, Muise now began an attack on the argument that intelligent design did not qualify as science. He cited several examples of phenomena that were once believed to have supernatural origins—the sun, for example—but were later understood by science. This seemed a fruitless and self-defeating argument: wasn't this precisely the plaintiffs' point?

He tried to argue for the scientific validity of intelligent design, but this too went nowhere because Miller was so much better informed. Muise might be able to articulate an aspect of intelligent design, but when Miller produced evidence refuting its detailed assertions, Muise had neither the language nor the wider scientific knowledge to be able to contradict him.

Now, however, Muise laid a trap that worked. It began when he pushed Miller to define his argument that intelligent design was a religious proposition rather than a scientific one.

"Is it your understanding of the theory of intelligent design that it requires the action of a supernatural power?"

As he had already been asked this question in several different forms, Miller became impatient.

"Okay. Again, intelligent design as I understand it presupposes that some features of living things are too complex to have been produced by evolution and therefore—and here's the answer to your question—they must be the product of an intelligent designer acting outside of nature."

So in his view, Muise probed, intelligent design required supernatural action?

"Perhaps it would be useful ... to define what supernatural means. The word 'super' means above. The word 'natural' of course means natural. The actions of an intelligent designer ... lie outside of scientific investigation. That means to me that it lies above, 'super', natural law ... Perhaps you could explain to me how an intelligent designer could act undetectably, outside of nature, to create order that evolution and natural law cannot, and not be supernatural?"

Muise accepted this lecture on the definition of supernatural, and then asked, "Do you know who Francis Crick is?"

Miller did, of course. Crick was a co-discoverer of the double-helical structure of DNA, sometimes referred to as the secret of life. One of the most celebrated scientists of all time, he received the Nobel Prize for medicine in 1962, along with James Watson and Maurice Wilkins.

Had Miller ever heard of a theory Crick advanced called directed panspermia?

He had. Crick suggested that the seeds of life on earth could have been purposely disseminated by an advanced extraterrestrial civilization. He and his co-writer, Leslie Orgel, speculated this might have been done by a civilization facing extinction who decided that the most effective way to enable life to continue elsewhere was to send out a scattershot of DNA. Perhaps they were attempting to "terraform" a planet—to turn it from one that had no life on it into one that could sustain life—so they could move to it later.

"And that was a hypothesis put forward by a Nobel laureate?"

"That's correct, sir."

"Is that a scientific claim?"

"Well, the specifics that Dr. Crick made is a scientific claim, because although it's not immediately a testable claim, it is a potentially testable claim in terms of if we are able to explore larger and larger fractions of the known universe, we may eventually find out if there is life in other places that could have been directed towards us. So it's a scientific claim in the sense that it's potentially testable."

"Is it a supernatural claim?"

"That's an interesting point, and in this particular case no, I would not regard that as a supernatural claim."

"So the fact that life-forms may have come from an intelligent being from another planet to this earth . . . that is not a supernatural explanation for a natural phenomenon?"

"It certainly is a far-fetched claim in that many scientists would point out that there's no evidence for it, but as Crick framed it, it certainly would be a claim, as I said, that is potentially testable and therefore would accord to natural law."

If I had been Muise, I would have gone on to ask: "If the idea of aliens spawning life on earth is a potentially testable proposition, why is it not also a potentially testable proposition to suggest that one day we might find aliens so advanced they created the entire universe?" One hypothesis might be more far-fetched than the other, but there was no qualitative difference, so why shouldn't the same rules apply?

Unfortunately, he slid off into other matters and didn't sink his teeth into his next good argument for so long that any connection between the two was lost.

Eventually, however, he asked Miller whether it was true to say that the ordinary meaning of a creationist was simply "any person who believes in an act of creation."

Miller agreed.

"And you believe that the universe was created by God?"

"I believe that God is the author of all things seen and unseen. So the answer to that, sir, is yes."

"In a sense that would make you a creationist using the definition—"

For the first time, Miller interrupted and then stumbled: "In the, as I think you and I discussed during the deposition, in that sense any person who is a theist, any person who accepts a supreme being, is a creationist in the ordinary meaning of the word because they believe in some sort of a creation event."

"And that would include yourself?"

"That would certainly include me."

"Now when you read the book of Genesis, you take that to be a spiritually correct account of the origins of our species, correct?"

"I take all of the Bible, including the Book of Job, the Book of Psalms, New Testament, and Genesis to be spiritually correct ... And the way in which I look at Genesis is that I see a series of commands of the Creator to the earth and its waters to bring forth life and, you know, without requiring—my church certainly doesn't—without requiring Genesis to be a literal history, you know, that's pretty much what happened, which is that the earth and its waters and so forth brought forth life."

Instead of closing the deal at this point, Muise once again meandered off into other issues, returning finally to his obsession with science being characterized by both continuity and change. His very last question was to ask whether this included Darwin's theory, to which Miller gave the predictable response.

It was an anticlimactic ending when it could have been so great. Had Muise put his two powerful arguments together at the very end, he could have gone out with something like this:

"So, let me just get this straight: You're a witness for the plaintiffs but in fact you are a creationist who believes that it's scientifically possible that all life on earth came about because of aliens?"

It would not have won the case, but it might have made Miller look a little ridiculous. For Muise to have said this, however, would have required him to poke fun at two concepts, one of which he was employed to defend, the other of which, like Miller, he truly believed in: God created the universe.

Other Primates

Five lawyers have been introduced: Eric Rothschild, Steve Harvey, and Vic Walczak for the plaintiffs, and Patrick Gillen and Robert Muise for the defendants, but this does not at all paint an accurate picture of all the primates doing battle.

Other lawyers from Pepper Hamilton came in, among them Thomas Schmidt, a gray-bearded, unquestionably decent man, whose function, it seemed, was to cross-examine defense witnesses who were either shy or so clearly feeble-minded or old that the sharp-elbowed style of the other three might actually render them unconscious.

Lending intellectual heft to this legal phalanx were Richard Katskee and Nick Matzke.

Katskee was from Americans United for Separation of Church and State, a party to the suit, while Matzke offered scientific expertise on behalf of the National Center for Science Education.

Katzkee, a tall, handsome man, soft spoken and civilized, was the scholar in the crowd, an expert on constitutional law, and the lawyer who would contribute the constitutional arguments in the final document briefing the judge on the plaintiffs' case. He also looked as if he had just stepped out of the pages of *GQ*.

Matzke, by contrast, was from Oakland, California, and usually looked as if he had just rolled out of bed. The NCSE staffer initially assigned to the Dover flare-up, he now briefed the lawyers on the arcane ins and outs of science. Bespectacled, in his thirties, he was tall and large and peered down at you with a look of beleaguered doubt, as if to say, "You're asking me this question about science, but you know and I know that you're not going to understand my answer, so, although I find this stuff fascinating, wouldn't you really rather go for a beer?"

Nick's boss, Eugenie Scott, came in and out while attending to her many obligations in defense of science education.

But the plaintiffs' team did not end here. Immediately behind the large table at which Eric, Vic, and Steve sat were two legal assistants, Kate Henson and Hedya Aryani, young women in a state of constant exhaustion, who produced the necessary ingredients for the lawyers' case. Both were attractive, lively, funny and intelligent, and to Hedya, the case had a special and personal significance. She was a victim of religious persecution, an Iranian of the Baha'i faith, whose family had been driven from Iran after the Ayatollah Khomeini took over in 1979.

At the end of the table, to one side of the screen, sat the unsung hero of the plaintiffs' case, Matthew McElvenny, technology specialist, the faultless computer genius whose computer held all the necessary exhibits—drawings of bacteria, excerpts from books and articles, depositions, even news video—and projected them on the screen. As anyone will tell you who has covered a trial, sleep is the most insidious enemy. From childhood on, the endless drone of the human voice has proved as effective as any narcotic, but when the Wizard of Oz was on, highlighting and scrolling without a single mistake, one inevitably perked up.

All these characters lived together in three rented apartments in the same building. Eric and Steve shared one, Kate and Hedya shared another, and Vic and Nick occupied the third. They worked long hours and usually ate together. Many mornings they walked to work as a team, Walczak, a Springsteen fan, often singing "Part Man, Part Monkey" as they approached the courthouse.

With Rothschild, Harvey, and Walczak in the lead, here was a team of eight to ten highly skilled professionals operating in an atmosphere of frictionless amiability. It was very clear—and this was another thing that might keep one awake—that these people liked each other and were having fun.

On the defense side, it was another matter. If the plaintiffs' legal team was a well-oiled collegial machine, the defense was a dysfunctional family with a frequently absent father.

The star lawyer—if not the lead lawyer—for the defendants, Richard Thompson bore an uncanny resemblance, at least in profile, to William Jennings Bryan. Bald, with strong features and a dark complexion, he had restless eyes and a body that was usually in motion. He had about him the air of a salesman.

Before forming the Thomas More Law Center, he was, between 1989 and 1996, the prosecuting attorney for Oakland County in Michigan, directing a staff of 180. He successfully defended the constitutionality of Michigan's mandatory life law for major drug dealers before the U.S. Supreme Court and created the state's first child sexual assault crimes unit.

But what made him famous was his almost obsessive attempt to send Dr. Jack Kevorkian, otherwise known as Dr. Death, to prison. After two high-profile, high-cost, and no-result prosecutions, the voters threw him out of office. He then formed the Thomas More Law Center, of which he was president and chief counsel with the help of arch-conservative Domino's Pizza magnate Tom Monaghan, a Catholic. The board chairman of TMLC was Bowie Kuhn, the former baseball commissioner.

Unlike his co-counsels, Gillen and Muise, Thompson, perhaps too busy pursuing Dr. Death to concentrate on procreation, had sired only three children. Clearly used to dealing with the press—he had appeared on *Larry King Live*, *The O'Reilly Factor*, and *The Joan Rivers Show*, to name but a few—Thompson was amiable, bombastic, and self-contradictory. While talking about gay marriage in Massachusetts he would say, "Judicial activism is destroying our culture," but he also described the Thomas More Law Center's purpose as being "to change the culture" through the courts.

He cross-examined witnesses in an oddly combative style, often turning toward the jury box (filled with an unsympathetic "jury" of reporters), before turning back to point his finger at the witness to ask a question whose substance frequently seemed to bear no relation to the tone in which it was asked. He would then sit down and rock back and forth in his chair, often staring up at the ceiling as if contemplating weightier matters.

Just as there was disjunction in the tone of his cross-examination, so it was with his presence or absence. It was hard to imagine that any case in the history of the Thomas More Law Center had ever been as important as this one. For the first few days, Thompson attended court dutifully. And then he disappeared. And then he came back. And then he left again. He missed several of the most significant witnesses and was present for some of the least. He was not even in court for the closing arguments. Perhaps as a longtime prosecutor he found the role of defense lawyer unnatural.

The most sinister of the Thomas More lawyers was Edward White, III. Ed, who came in only to question a couple of witnesses, was famous for defending antiabortion activists who listed doctors' names and addresses on an Internet site named the Nuremberg Files. These lists took the form of Western-style "wanted" posters, and three doctors on the site were duly killed in the nineties, after which X's were pasted in congratulatory fashion over their faces. (The site has been shut down, but if you search the web for Christians of this violent persuasion you can still find lists with the names of the three murdered doctors—and two law enforcement officers—now struck through with a tasteful line.)

Ed's face was not one within which I could find anything to like. In repose, his head was tilted back in petulant defiance. A superior sneer worked his mouth, and his eyes were arrogant and cold. But he was rarely in repose. Every few minutes, his hand would reach up to scratch his nose, then readjust his watch, his glasses, the knot of his tie; now a jacket-shrug, a chin-scratch, a neck-scratch, then back to the glasses, the tie—and this cycle would repeat two or three times before he settled.

For a while there was a junior lawyer, but she went the way of Thompson. Sometimes it was ten against one, Patrick Gillen alone, smiling dutifully, fumbling for his own documents. By prior arrangement—or out of simple human decency—the plaintiffs' machinery was often put at his disposal so he could show his documents on screen.

The impression I had was that Pizza Man needed to step forward with a new infusion of cash.

. . . .

As Margaret Talbot wrote in her *New Yorker* piece about the trial, Judge John Jones had "the rugged charm of a nineteen-forties movie star; he sounded and looked like a cross between Robert Mitchum and William Holden."

A Republican appointed to the bench by George W. Bush, he had grown up near the small town of Pottsville, an hour or so north of Harrisburg, and had been educated entirely in Pennsylvania. He went to Blue Mountain High School, a public school, and then on to Mercersburg Academy, a prep school whose graduates included Jimmy Stewart and Benicio Del Toro. Though he was as handsome as the latter, his character was more like Jimmy Stewart's. Although, as he put it, he "was a child of the sixties" and was the president of his fraternity at college ("and that was back in the days of Animal House"), he took law school seriously.

He graduated from the Dickinson School of Law, was admitted to the bar in December 1980 on the day John Lennon was shot, and went to work for a small law firm in Pottsville while also clerking for a judge. Before long, he became an assistant public defender before starting his own law firm in 1986. During these years he was involved in cases ranging from petty theft to homicide. A year after taking his bar exam, he met and married his wife, Beth, a local girl, daughter of a man who sold hearing aids. She was a teacher.

Jones was a civilized and thoughtful man. More than just polite, he was courteous, a gentleman, a man who treated everyone around him with equal respect. But he was also funny. One day, an objection was raised as to the admissibility of a question. A long and acrimonious debate followed, with lawyers on both sides giving it their all. When Jones finally ruled the question legitimate, it was put to the witness again.

"I don't know," he replied.

"After all that," said Jones.

When I complemented him on his humor, he said he hoped it helped relax tension, but he was never cruel. He had experienced cruelty from the bench when he was a defense lawyer and was determined never to abuse his power in this way.

In 1992, he ran for Congress as a Republican and lost the race narrowly. The loss affected him deeply. In 2002, he was appointed chairman of Pennsylvania's State Liquor Control Board. One of the largest state monopolies in America, if not the largest, it bought, sold, and regulated alcohol in the state. In his day, it sold over a billion dollars of alcohol a year.

"I went," he said, "from being a small-town lawyer to being the de facto CEO of a Fortune 500 company." Unlike most CEOs, he also had to testify in budgetary hearings before the state's House and Senate and therefore knew his way around politics.

During the trial many people, impressed by his decency, intelligence, and tolerance, remarked that he would make a far better president than the man who had appointed him.

The Teachers' Story

During the trial the other reporters and I occasionally drove over to Dover High to try to extract some juicy comments from the local kids or to talk to one or another of the teachers. These attempts were usually fruitless because by this time, everyone at the school was sick of us, and the teachers had, I think, been told not to talk.

As I was not just writing about the trial but also directing a documentary about the larger issues raised by the case, I was always in search of an illuminating on-camera response.

On one occasion, having been forcefully told to get

off school property, I found myself whistling from the sidewalk to five teenaged boys skateboarding in the parking lot that lay between me and the front entrance. Eventually they came over and agreed to talk. I asked them what their religious beliefs were. They were all atheists. So much for that.

Another time I got two tenth-grade girls to talk on camera. One furtively smoked a cigarette, turning away from my lens to puff and exhale as if that would make a difference were I ever to show the footage. Neither knew anything about anything. In fact, this was the only interesting thing about them: their total lack of interest. They knew nothing about either evolution or intelligent design, though one had a dim memory of hearing about one or the other of them in some long-ago class. One said her mother didn't vote and she wasn't going to either.

But if she didn't vote, I asked, how could she complain if she or a brother or a boyfriend got sent off to war? She told me such things didn't happen to people like her, or to her family or friends, and she didn't care about anyone else.

Whenever I spoke about the school, one name always came up: Bert Spahr.

Bert—a contraction of Bertha—had been a teacher at Dover for forty-one years and was now the science department chair. A legend in the district, she taught chemistry and was on the verge of teaching some of the grandchildren of former students. Having been there so long, she had taught a large proportion of the people I met, including school board member Jeff Brown. ("Oh, my lord, what a character," she said of him when I eventually got to talk to her.) No one forgot the experience of being taught by Bert, and many went on to careers in the sciences.

Some months after the trial, I went back to Dover and sat down to talk with her and two fellow science teachers, Jen Miller, the head of the biology department, and Robert Eshbach. The meeting took place in Bert's classroom, a large, handsome room containing sturdy lab benches, a periodic table on the wall, and various gifts from students placed on a shelf.

Rob, Jen, and I sat on too-small school chairs while Bert sat dwarfed behind a massive wooden block of a desk.

A short woman with shining brown eyes and a throaty laugh, Bert wore a colorful diamond-checked sweater and equally colorful slacks. Her father was an Italian immigrant and farmer, but she preferred to think of herself not as Italian but Sicilian. As she would say to students when they misbehaved, "How would you like to see a short angry Sicilian flying across this desk toward you?!"

Jen and Rob were both children of ministers. Jen was in her thirties and conservatively dressed, at least compared with Bert. An attractive woman with dark curly hair, she was married to a fellow teacher at the school, a history teacher, and was the mother of young children. She was shy (again perhaps only in comparison with Bert) but shared Bert's humorous outlook.

Rob, father of three young children, was good looking, bearded, and hawklike in appearance but gentle and thoughtful in manner. He had gone to Dover High, in fact had studied under Bert, and considered it an honor and a pleasure to be working with her now. All three, though obviously disturbed by what had happened, laughed frequently about it, most often at Bert's pugnacity.

The science department had been one of the most stable in the school. Jen had taught twelve years here. Another biology teacher, Robert Linker, had taught a similar length of time. They had had little contact with either the board or the administration. They simply got on with teaching to the state standards, occasionally pressing for newer books or better facilities, but never encountering much resistance or conflict.

Before intelligent design intruded, Robert Linker, would, when approaching evolution, draw a line down the center of the chalkboard, write "creationism" on one side and "evolution" on the other, and then explain that he was not qualified to talk about the former so would be concentrating only on the latter. He would also show videos, one of which, *Apes to Man,* explained that man, while sharing a common ancestor with the ape, was not actually descended directly from him.

Jen Miller, meanwhile, in order to show the enormous amount of time involved in evolution, took her students out into the hallway and laid out a long measuring tape so she and her students could mark in the approximate time at which certain species are thought to have appeared.

That evolution was a potentially contentious issue was, of course, something they were well aware of. It was clearly on Bert's and Jen's minds when Robert Eshbach applied for the job. He recalled being rather surprised when they asked him whether as a minister's son he had any problems with teaching evolution.

"We always had fundamentalist Christians in our community," Bert told me, "and we were always very careful when we got to this area . . . I'm not saying we soft-shoed it, but we certainly tried to be the least offensive that we could be." Alan Bonsell's father, Donald, who was on the school board at one time, wanted creationism taught, but the principal at that time was a biology teacher and nipped the whole thing in the bud.

The first time Bert began to suspect the subject might be about to reappear, perhaps in a more virulent form, was in the fall of 2002, when a mural disappeared. Painted by a student in the late nineties, it was sixteen feet by four and rested on the chalk board tray in Room 217. It depicted the evolution of man, with an ape at one end, crouching, and a man at the other, standing. Unlike other art donated by students, it had never been nailed to a wall in the hallway "because the gentleman who was at that point head of buildings and grounds was offended by what it depicted . . . it was a beautiful piece of artwork."

A few days before the beginning of school, the teachers came in to prepare. The teacher to whom the mural had been given had just left the school, and Rob Eshbach was about to move into his classroom. "My room," said Bert, "was connected to his room and there was a closet between us, so I could go through the closet and go into his room, and I was over there seeing if he had the books, the pencils, the paper, whatever, and I turn around—it's gone. Now it would be tough not to know it was gone, and I of course said to Rob, 'Do you know anything about this mural?' Of course, he didn't know what I was talking about. I went out to the janitors who service our building, and I said, 'What do you know about this mural?' knowing that there were people who did not approve of its content. Well, no one knew anything about it."

It had disappeared between a Friday and a Monday. "So then I went down to the assistant principal, and I said, 'I think you need to go in search of what happened to my mural.'"

The assistant principal went in search of it and could not find it. Bert suggested that he ask the head of buildings and grounds, Mr. Reeser, if he knew anything. When asked, he "admitted that he had removed the mural from the room over the weekend—and burned it . . . Now that was when I had kind of an inkling that we were in trouble."

The man was not punished for what Bert thought of as theft and destruction of school property, and that made her even more nervous. Two years later, when she heard that Buckingham had shown a picture of the mural to someone at a school board meeting, she asked him where he had gotten it. "He wouldn't answer me. Well, the only way he could have obtained it is if he had been present right before it went up in flames." When Jen Miller pushed the point, Buckingham admitted that he had "gleefully watched it burn."

I asked what Buckingham was like. One of the three—I won't say which—simply said, "Jackass." They described his demeanor as rude and bullying, but Bert believed that although his was often the loudest voice, Alan Bonsell, whose manner was less confrontational but more patronizing, was actually the persistent driving force.

As for school board member Sheila Harkins, no one could figure her out. She was a Quaker, a member of a faith with no objection to evolution. Whenever she was running the board meetings and didn't like something, she'd simply bang the gavel and call for a recess, leaving everyone wondering how long they'd have to sit and wait for a resolution to whatever problem was being dealt with.

"Such screaming matches—oh, my glory, it was the most mortifying thing," said Bert. "There was not to my knowledge any one of them that had any background in science training, and yet there they sat, some without even a high school education, telling us how we are going to run a science curriculum to educate our children."

One day in the spring of 2003, the assistant superintendent, Mike Baksa, came to visit Bert and "gave me the red flag that said there is a board member who wants to introduce creationism alongside of the teaching of evolution ... That happened one evening after school. He'd always come—bless his heart—at ten after three, which drove me crazy." When she asked who the board member was, she was told it was Alan Bonsell. "The next morning I immediately went to my supervisor and said, 'I think we better get prepared, because I'm afraid of what is coming.'"

In the following months, Baksa occasionally dropped by, usually at lunch time, and continue to talk to the science teachers about the board and its growing desire to "balance" evolution with creationism. He was, of course, rebuffed, but he kept coming back to float the same idea again and again.

"We did attempt to clarify this issue to him," said Bert, "but I believe his background is not in science."

In the fall of 2003, Alan Bonsell's son was in the ninth grade and would soon be studying evolution. As Bonsell was clearly not going to stop pestering Baksa, and as Baksa was clearly not going to stop pestering the teachers, the teachers eventually suggested they meet with Alan in person. By this time, they had learned from Baksa that Alan was a young-earth creationist (that is to say he believed the earth was younger than ten thousand years), so they knew what to expect.

As the senior biology teacher, Jen Miller was chosen to be the principal spokesperson at the meeting, but most if not all of the science department came too. Bonsell was concerned, Jen explained, "that we would convey something to the students that was in opposition to what their parents were conveying to them at home and didn't want to put the teachers in the middle of, you know, having the students say, well, someone is lying basically."

She explained to him how the biology teachers taught evolution and that—this being the obvious problem—they taught origin of *species*, not origin of *life*. In other words, the teaching of evolution—which proposes that all life on earth, including man, is descended from some single-cell organism billions of years ago, as opposed to the biblical version wherein he arrives fully formed and at the center of God's recent creation—had already been toned down to pander to the creationist sensibilities of the locals.

Robert Linker remembered that Jen had to explain everything to Bonsell several times before he understood. After which, "if I remember correctly," said Linker, "he looked at me and said, 'Is that how the biology teachers teach it?'" to which Linker replied that yes indeed it was.

During the trial, when Bert was asked, "Did you speak much at the meeting?" she replied, "Not as much as I usually do." The teachers left the meeting believing that Bonsell had more or less grasped the distinction. "We felt it was a very congenial meeting," said Bert.

It may have been congenial, but it was clearly a warning, and both Robert Linker and Jen Miller changed the way they taught. Linker stopped showing videos about evolution, and Jen stopped laying out her evolution timeline in the hallway.

The next thing the teachers feared was an attack on health education. Teen pregnancy has always been a problem in the district—I was astonished by how many women I met had kids before graduating from high school—and at the time of my second visit to Dover, ten to twelve girls in the senior class were pregnant.

The situation was so bad the school had bought an infant-sized doll from the "Baby Think It Over" program, a doll that cried and demanded feeding and attention so girls could take it home and experience the realities of motherhood. In some places (I don't know whether this was the case in Dover), the doll is actually tethered to the mother. The details of how well the girl treats the child—including evidence of shaken baby syndrome—are recorded by the device and can be viewed by teachers.

In either late 2003 or early 2004, Alan, now president of the board, selected a new chair for the curriculum committee. He could have chosen Carol Brown, with her degree in secondary education. Instead he chose

Bill Buckingham. When the science teachers renewed their requests for new textbooks, they were asked to submit some of the old ones and the new ones, just for comparison.

The teachers met several times with the new curriculum committee, whose other two members were now Carol Brown and Sheila Harkins. The fiscally conservative Sheila questioned the need for some of the new books.

Buckingham remained fairly quiet, at least at first, but soon began to express his distaste for the Miller and Levine book, *Biology*, because of its treatment of evolution.

. . . .

Plaintiff Bryan Rehm had been a physics teacher on the science faculty at Dover during much of the controversy, but eventually he left to go teach in another school. He was a tall, large, self-contained man in his late twenties, and I have to admit that I did not like him at first. He had a slightly superior air, and his notion of being communicative was to permit himself the occasional shrug or grunt of assent or disagreement without benefit of a smile.

As if to compensate for his taciturnity, his wife, Christie, also a plaintiff and also a teacher, was an effusive, unambiguously friendly woman with a broad smile and warm, humorous eyes. She had recently given birth to their fourth child.

It was not until long after the trial was over, when I visited Bryan at his home and saw him helping his daughter do her homework that I came to understand and admire him. Bryan was a natural-born teacher and a true believer in education, and if he was inexpressive, it was in part because he was trying to contain his rage about what had happened. He was deeply offended by this attack because what was being attacked was a practice he and Christie considered their calling rather than their job.

The Rehms were involved in more activities than seemed humanly possible—even if one did not have so many children—so they were unable to see me until eight o'clock one weekday night. Even then, Christie was still at her school and came home only later.

Bryan and I sat at the kitchen table chatting as his daughter worked diligently on her homework. Bryan's mother was a teacher, his father a factory worker. He was raised in the Lutheran church, drifted away from it in college, and then, while working on a Habitat for Humanity project in West Virginia, found that most of his co-workers belonged to the United Church of Christ, or UCC. When he met Christie at his first teaching job and found she too was a member of the UCC, he became fully involved.

Having considered a music career in college—studio sound recording—and he played the guitar, he soon became co-director of the children's choir at the local UCC church. I asked him what percentage of his friends came from the church, and, after a moment's thought, he reckoned it was about half. His and Christie's most recent friends came out of the case.

The Lutheran church he attended as a child was more reflective than Evangelical and certainly was not fundamentalist. When abortion was discussed, it was in a tone of intellectual inquiry, an exploration of ideas about when life began rather than an exercise in moral condemnation. There were gay parishioners, though none were flamboyantly out, and there was no church policy on the issue.

The UCC was more explicit. "The UCC has taken a lot of heat in the past year," Bryan told me, "because at their national conference they decided that it would be best to be the open and affirming church, so everybody is welcome, we don't care what your status is . . . we think it's important you come and hear the word of God, so our doors are open: come in, have a seat, and we'd like to get to know you."

His mention of an open door policy was in interesting contrast to the religion Alan Bonsell subscribed to at the Church of the Open Door in York, because if their website was any guide, the door that concerned them most was the trapdoor to hell.

Bryan found Alan's certitude amusing at first. He would come home and say to Christie, "You're not going to believe what happened today!"

but after a while it became alarming. Polite at first, Alan became increasingly arrogant and indifferent to their arguments. If it was suggested that I.D. might not be science, he would say things like, "Well, you need to watch Kent Hovind's video *Evolution Is Stupid*," or "We've got this think tank that says it's science, so we know it's science," or that he knew it was science because he'd "read it in a magazine." When questioned as to which magazine, Alan smirked and said he couldn't remember.

At a certain point it seemed to Bryan that Alan had given up caring what anyone thought. He had the power. He was right. That was it.

Christie came bustling in. She laughingly apologized for not having arrived sooner. She taught at the same school as Bryan, and "we're doing a food drive and it's a competition between homerooms to see who can get the most food and my homeroom collected money this morning and gave me this big pile of money and said, 'Go buy us ramen noodles,' because they know ramen noodles are ten cents a pack! So I have a car filled with ramen noodles!"

Wherever I went in Dover, I found this charitable impulse. Barrie Callahan's daughter, Katie, told me about "Thon," a dance marathon where the students danced through a whole night in order to raise money for "The Four Diamonds Fund, Conquering Childhood Cancer" based at Penn State Hospital in nearby Hershey.

Christie's mother got pregnant while in her senior year of high school. She married the boy who got her pregnant, but both were fairly wild, and the marriage lasted only a few years. Having a father who was only sixteen years older than herself was not really like having a father at all, Christie told me; it was more like having an older brother. When her mother married her stepfather, things began to settle down a little, but Christie always felt unwanted in a way, "like I wasn't useful."

As is so often the case, she followed in her mother's footsteps and also got pregnant while still at school. "You want certain things, and you don't know how to quite get them, and you're not feeling good where you are, and you search for any way to get out, to figure it out, and I'm not saying that it was an intentional act to get pregnant so that I could have somebody that loved me," but it was almost certainly an underlying cause. The relationship lasted until her daughter was about two.

She talked about her mother and father a little wistfully but without blame or rancor. The life she led as a child gave her direction: from an early age she knew she wanted to be a better parent than her young mother had been capable of.

In spite of having a child, Christie went to college and got a degree in English and journalism. For a while she did freelance work for a local newspaper; then she went back to school to get her teaching certification. She had been teaching since 1999.

You could see how Bryan, and even more Christie—who was the first in her family to go to college and who had, in spite of encumbrances that would have crushed a lesser person, achieved so much through education—would find it hard to understand the attitudes of men like Buckingham and Bonsell.

They puzzled over their motives and their belligerence. "It's almost like the ability to empathize with somebody, or sympathize with them, is beyond their grasp," said Bryan. "They are totally, it seems, incapable of putting themselves in anybody else's shoes. Their base of experience is so narrow and centered on themselves that it's almost like there's no imagination there for what it would be like to be any other way."

A symbol of that insularity can be found on the website of the church where Buckingham worshipped, Harmony Grove, which directs parishioners to businesses owned by believers of their own type. (There is in fact a Christian Yellow Pages, which makes it possible for someone to go to almost any place in America and practice this form of economic favoritism.)

"I just think the irony is so deep here, you know, in so many areas," added Christie. "The fact that we have a school board of people who really hate education is just beyond belief. It's absolutely insane. What's their reason for really wanting to be on the school board? To control the school, to make the changes that they see fit, to really destroy public education; that's their goal."

The atmosphere in the school became intolerable for Bryan, and it bled into life outside the school. If you went out to dinner, you worried about who might sit down next to you and what they might be thinking. In spite of all his and Christie's Christian activities (including running their church's Vacation Bible School), Bryan was accused of being an atheist. His daughter, who was in the same class as Alan Bonsell's daughter, was asked by other students whether she believed she came from monkeys and why her parents were causing all this trouble.

No matter that the fundamentalists on the school board were now in the majority, they still behaved like victims, as if they were the ones being persecuted for their beliefs even as they imposed them on others.

Toward the end of my interview with the Rehms, Christie said, "We've been told by this group of people that we're not the right kind of Christians, you know, we're not actually Christians, we're something 'other,' and I think, well actually I don't mind being something other because if Christianity is what you are, then I'm not a Christian. I'm not."

As I left the house I couldn't help turning her remark in another direction. If I were a Christian, these were the kinds of Christians I would want to be like.

The Audible Ping

THE AUDIBLE PING I mentioned so many pages ago was heard at the first of two school board meetings in June of 2004, and it would become a clang at the following meeting. Unsurprisingly, it was Barrie Callahan who set the whole thing off.

She had become intensely frustrated by the board's refusal to buy the biology book she felt her daughter needed. Looking at the agenda for the June 7 meeting of the school board, she noticed that there was no mention of the biology textbook. "So I felt," said Callahan, "that I just had to approach the board one more time and ask them why the biology books were not scheduled for approval."

Buckingham told her that it was because the biology book was "laced with Darwinism."

"So, this is about evolution," she said, throwing up her hands, and sat down.

Max Pell, a student who had graduated with her son, was sitting at the same table. He had realized in his senior year that he liked helping people and, with medical school in mind, was now studying biology at Penn State. He asked Callahan if she thought it would be okay if he spoke. His girlfriend, Emily, was the student representative on the board and was sitting at the table with the board members. He did not want to embarrass her, but he felt compelled to raise objections. Callahan encouraged him to speak his mind. He stood up and headed for the podium.

"At that point," Max told me, "it wasn't even such a full-blown thing. They were just refusing the science textbook because it was, quote, 'laced with Darwinism,' which apparently was a bad thing."

Arriving at the podium, Max explained to the board that evolution was pretty much a working law in biology and that the board would be disenfranchising the students by taking it out of the curriculum. Buckingham was condescending, as if he was enlightened and Max was not. But Max persisted. Staying as calm as he could, he went on trying to convey the importance of evolution. They argued for a while until Buckingham, who seemed to be getting frustrated, said:

"Well, you're a perfect example of what happens to students when they go to college! They get brainwashed!"

Max laughed when he told me this, but at the time it shocked him, as it shocked many people who heard it. "It was so insulting, and I had never been spoken to like that before—I mean in public—like maybe by family, you know, but . . ."

Noel Wenrich, another creationist on the board, told Max that a scientific theory became a theory by repetition. "In other words, if you just keep repeating it and repeating it and repeating it, whatever it is, that's how science becomes a theory."

Max remained at the podium. When he suggested that the board might get into legal trouble if the school tried to teach creationism, Bonsell offered that there were only two theories of evolution, Darwin's theory and creationism, and so long as both were taught, there could not be a problem.

Buckingham didn't seem to care whether there was a problem or not. As far as he was concerned, he bullishly asserted, the whole concept of the separation of church and state was a "myth," and he was ready for whatever battle might ensue. After a while, he stood up, went over to Alan Bonsell, and showed him something. He then turned to Max and, showing him the photograph of the mural that had been destroyed, said, "You're not going to tell me that I came from apes, and if you insist on it, which side of your family came from apes?"

At the end of the meeting, the board stated it was not going to buy *Biology* but would instead start looking for a new book that included the teaching of creationism. Buckingham and Bonsell would later claim this was not true and that they had spoken only of looking for one that included intelligent design.

Two local reporters who would figure heavily in the trial, Heidi Bernhard-Bubb of the *York Dispatch* and Joe Maldonado of the *York Daily Record*, were covering the meeting for their respective papers.

When it was over, Maldonado spoke to Buckingham, who told him all he wanted was a book that offered a balance between Christian views of creation and evolution. "There needn't be consideration of the beliefs of Hindus, Buddhists, Muslims, or other faiths and views," he went on to say. "This country wasn't founded on Muslim beliefs or evolution, this country was founded on Christianity, and our students should be taught as such."

Christie Rehm, who attended the meeting with Bryan, overheard the remark.

"I remember talking to my husband about it on the way home, because we're both teachers, and when I hear things like that, I immediately think of my students, and I was thinking about the diverse group of students that I have in my classroom, who all have different religious viewpoints, and how difficult it would be to tell one student that, you know, we can't express your belief but we can express that person's belief in the classroom."

Heidi Bernhard-Bubb went home and called the ACLU in nearby Harrisburg. Staff attorney Paula Knudsen told her if anyone tried to put a textbook into the curriculum that taught creationism it could lead to a federal court case. Heidi reported this in an article that was published the following day, but if Buckingham read the article, he was not deterred. Over the next few days, he started looking for a book that would suit him and his fellow fundamentalists.

What he told his wife, Charlotte, about the meeting will never be known; but whatever it was, she took it to heart and would have plenty to say about it at the next meeting.

. . . .

The story of the June 7 meeting had been picked up by the wider media. Buckingham's fellow board member, Jeff Brown, remembered that he ran into Buckingham, who "was fairly glowing because he was getting letters of support from all over the country . . . But the thing that really impressed him most was an institution in Seattle."

The Discovery Institute, a think tank in Seattle, Washington, about which I'll have more to say later, was the most important organization promoting intelligent design. It is not known exactly what transpired between that organization and Buckingham, but one can be fairly sure that Seth Cooper, the attorney with the group who had called Buckingham, would have advised him not to use the "C" word—creationism. The main strategy of the Discovery Institute was to escape creationism and replace it with intelligent design. They were, in fact, so conscious of legal precedent and so wily that they did not even advocate teaching intelligent design in school any more but instead recommended that teachers "teach the controversy."

And yet—according to the reporters, and later to other witnesses—Buckingham continued with the "C" word long after he had spoken with Seth Cooper. There are two plausible explanations for this. Bill was suffering from the disorienting effects of OxyContin, or withdrawal from it, and simply forgot. Or Bill was a cantankerous guy who liked to mix it up. Or both.

The next meeting started off promisingly, with Buckingham apologizing for his behavior at the previous meeting. But the apology soon degenerated into his stating once again that he believed the separation of church and state was a myth and that "liberals in black robes" were taking away the rights of Christians.

"I must be who I am," he said, "and I'm not politically correct."

(This was undoubtedly one of his truer statements. At some point during one of these two meetings, Carol Brown remembers him saying that "anyone who does not agree with bringing faith into the schools is un-American and should return to the place from which he or she came.")

After Bill's "apology," his wife, Charlotte, went up to the podium and began to talk. Perhaps "talk" is the wrong word. Charlotte, who was a financial secretary at the Harmony Grove Church, where both she and her husband worshipped, had a gospel music ministry. Using accompaniment tapes, she sang Southern gospel hymns about eight to ten times a year at nursing homes, retirement centers, retreats, and family picnics, for which she received "free-will offerings."

She did not testify in court, but her deposition revealed that, like so many of the defendants, she had done little or no research into either evolution or intelligent design and had barely discussed it with Bill.

On this latter point she said, "I just want you to understand that my husband and I had very limited conversations, because he is not a great talker at home. And you know, basically, we didn't have a lot of discussions about any of this." None of this prevented her from having absolute confidence in her views.

"I always refer to that meeting," said Jeff Brown, "as our 'Come to Jesus Meeting,' because she put on a fifteen-minute tent show revival for our benefit. She called for the teaching of creationism in school, called us all to return to our roots in the Bible and to turn away from the world and its ways—and Darwin in particular. Normally public comment is limited to five minutes, but Alan Bonsell, who was the board president at that time, made no attempt to cut her short. She went a full fifteen minutes, and it seemed more like an hour and a half to me, because all I needed at that point—if I had had a handful of sawdust to hold under my nose, I could just close my eyes and I would've been in a revivalist tent . . . She is a professional evangelist, and she was giving us the whole show free of charge. She didn't even ask for a collection afterward. I was impressed."

Fred Callahan, probably the most measured and temperate of the witnesses, said, "It was a real religious polemic. It went on for, I would have guessed, for fifteen, twenty minutes. It was tantamount to a religious sermon, I would say."

As Charlotte spoke there were muttered "Amens" from the people sitting on either side of Carol Brown: one from Heather Geesey, the other from Bill Buckingham.

"After she got done speaking," Jeff said, "my wife and I began raising objections to teaching creationism, such as 'It's illegal. You cannot teach creationism. The Supreme Court has already ruled that you cannot teach creationism in school.'"

Pretty soon Jeff was banging his fist on the table. "And, at that point," Jeff said, "Mr. Buckingham looked me in the eye and said:

"'Two thousand years ago someone died on a cross for us! Isn't it time we take a stand for him?'"

As Christians, the Browns did not oppose creationism so long as it was not taught in science class. Carol suggested a better place for it would be in a class called Comparative World Religions, "so that our students could be introduced to the major world faiths and the way in which they're the same and the way in which they differ, in particular the fact that every major world religion has at its core what we Christians call the Golden Rule: 'Do unto others as you would have them do unto you.'"

To digress a moment, it has always interested me that one of the claims religion makes is that without its extensive texts, moral decisions would be almost impossible to make. But in truth, isn't the single beautiful concept of the Golden Rule almost sufficient on its own? Of course, there are many complex moral dilemmas in life, and sometimes you must make a choice that causes you to do unto another something you would not wish to have done to yourself. But I have read the bible from cover to cover and made some forays into the Koran, and both are so crammed with contradictory instructions (in the bible, for example, Thou Shalt Not Kill a few chapters from Stone The Homos to death), and both are so archaic, so drenched in blood, anger, and vengefulness that in aggregate, they actually constitute a massive impediment to simple decency, let alone morality.

However, to obey the simple dictates of the Golden Rule requires imagination, and that, as Bryan Rehm pointed out, is often what is most absent in the fundamentalist.

In *De Profundis,* Oscar Wilde wrote, "Christ's place is indeed with the poets. His whole conception of Humanity sprang right out of the imagination and can only be realized by it ... There is still something to me almost incredible in the idea of a young Galilean peasant imagining that he could bear on his own shoulders the burden of the entire world: all that had been already done and suffered and all that was yet to be done and suffered."

This vision of compassion brought forth by imagination stands in gruesome contrast to the shuttered certitude of men like Buckingham and Bonsell. What a fall from grace, from the poet of sympathy to the bullies of Dover—what an ugly, distorted, complicated, hampered, and limiting thing the whole affair has become.

Buckingham next accused Jeff of cowardice, saying he was "glad he had not been fighting during the American Revolution because we would still have a queen on the throne ruling our country."

Bert Spahr, who was attending with most of the science department, now went up to the podium. She spoke in support of the biology book, saying that its treatment of evolution was sensitive, that it was the most widely used book in the country, and that the number of pages devoted to evolution were minimal. In closing, she urged that the board remember the legal issues involved.

Buckingham responded, "Where did you get your law degree?"

"It was a very short, pointed barb," Fred Callahan said. "I'd have to categorize it as being a very gratuitous slap in the face. As I recall, there was an audible gasp from the crowd."

Warren Eshbach—retired pastor from the Church of the Brethren, and science teacher Robert Eshbach's father—expressed concern at how this issue was dividing the community. He pleaded with the board to find a compromise. Buckingham agreed to meet with the teachers to discuss the biology textbook. That was the best he could do.

Because of news reports of the previous school board meeting, a representative from Americans United for Separation of Church and State had come to this one. He too warned the board that if they went down the creationist road they'd almost certainly run into a lawsuit, but Bellicose Bill was not in a listening mood.

The next day, his outburst—"Two thousand years ago someone died on a cross for us! Isn't it time we take a stand for him?"—was widely reported in newspapers. It would become the most famous remark to emerge from Dover.

A Libertarian Buddhist Girl Scout Leader

PLAINTIFF BETH EVELAND was a large bustling woman. She was eccentric, intelligent, passionate, and had a good sense of humor.

When I went to visit her at her rambling house opposite Buckingham's church, it was well into a weekday evening, but she was still not home, so I sat and talked for a while with her husband. Bud was a quiet, good-looking guy with a beard, who worked in the plumbing and heating business. He was awaiting a hundred and thirty-two cases of Girl Scout cookies, which were to be delivered to Beth, a Girl Scout leader, for distribution the next day.

I heard the rest of the family before I saw them—the

bang of car doors, laughter, a thunderous clatter on the porch—and then in they came, Beth herding her two girls ahead of her, a six-year-old and a seven-year-old. All three were returning from a Brownie event and seemed more energized by it than drained.

After hurtling off to feed the family rabbit, a free-roaming creature, Beth slammed herself down into one of two big sofas. Bud went upstairs to put the kids to bed and then went to bed himself. The job he was currently working on was a ninety-minute drive away, requiring him to rise at five.

When I transcribed my interview, I realized that at least a third of it was taken up with Beth's laughter, laughter that came rolling out of her both during and after many sentences.

Beth was raised Catholic but her family left the church when it refused to bury two stillborn (and therefore unbaptized) children in its graveyard. She had a happy childhood, her family—unlike many others at that time—staying together. She was a "band geek" and fairly popular. Her father, a volunteer fireman, raised her to think she had a duty to her community, which is why she became involved in the Girl Scouts. With an associate's degree in "legal secretarial science" (she enunciated this with some grandeur, laughing), she worked for a local attorney, describing herself later as his "pit bull."

She was a smart and buoyant woman who made the best of everything. One of her daughters had an illness as an infant and Beth had to inject medicine into her abdomen for several weeks. This was extremely painful for the child and awful for the mother, but the alternative was to take her somewhere to have it done by a stranger. As she said in an aside, "Jump in front of a train for your kids, man—you gotta do what you gotta do."

At the same time as this, Bud's father came to the house suffering from stomach cancer, and she was looking after him too. "I'd help him take showers and get dressed and all that. It takes a lot out of you to care for somebody, let alone a family member, but I really had such satisfaction taking care of him." Because of this experience, she was now—on top of her job, motherhood, and her scouting activities—working toward a nursing degree.

She had moved to Dover more than a decade ago but still felt like an outsider, though, paradoxically, less so since the trial. Growing up in the nearby town of York, she thought of herself as a city girl and, in fact, remembered that she and her school friends had made fun of people in Dover, "'cause it's rural, they probably go cow-tipping, you know, they're farmers."

Even so, when she read the article in the paper about the first school board meeting in June, she was, if not entirely surprised, definitely shocked.

"I go 'You've got to be *kidding* me! In this day and age?!' I told Bud, I said: 'I gotta go to a school board meeting, I gotta see what's going on . . .' Bud was as appalled as I was, but he's more of a kinda just sit-back guy. I'm not, man, my mouth's too big, I gotta get out there and say 'Wait a minute!'"

She immediately wrote a letter to the local papers, which was published.

> As a parent in the Dover Area School District, I must convey my shock and utter dismay at William Buckingham's comments regarding the search for new biology texts for the high school. I am especially upset with Mr. Buckingham's comments as quoted in Wednesday's York Daily Record: "This country wasn't founded on Muslim beliefs or evolution. This country was founded on Christianity, and our students should be taught as such." This statement is in direct contradiction to the mission statement of the Dover schools.
>
> "In partnership with family and community to educate students, we emphasize sound, basic skills and nurture the diverse needs of our students as they strive to become lifelong learners and contributing members of our global society." What a slap in the face to many of the parents and taxpayers of the Dover area. How sad that a member of our own school board would be so closed-minded and not want to carry on the mission of Dover schools.
>
> His ignorance will not only hold back children attending Dover area schools, but also reinforce other communities' views that

> Dover is a backwards, closed-minded community. If it was simply a matter of selecting a text that gives two contradicting scientific theories equal time, that would be an entirely different matter, but it's not. Creationism is religion, plain and simple.
>
> Buckingham's comments offend me, not because they are religious in nature, but because it is my duty to teach my children about religion as I see fit, not the Dover Area School District during a biology class.

When the second meeting occurred on June 14, she was there.

"What I saw there ... was just crazy, people calling out other people about their patriotism ... When Buckingham made the comment about 'liberals in black robes,' I'm sitting in the front row, and I'm thinking: 'Am I on *Mars*?' You know, 'Are these other people sitting here hearing what I'm hearing!?'"

Buckingham, she said, was indignant about almost everything. Like me, Beth had read Charlotte Buckingham's deposition and been struck by her remark about there not being a lot of conversation in the house. "He'd have a frying pan sticking out the side of his head if I was living with him," she said, laughing even harder than usual, "'cause I'd just *have* to club him, I would, I just would, I'm sorry!"

The more meetings she went to, the worse it got. It "started getting crazier and crazier, you know; Noel Wenrich was this kind of loose cannon, and him and Buckingham, one minute they'd be like buddies voting together and the next minute Buckingham's questioning his patriotism and Wenrich's like throwing chairs and—*literally*—and 'Let's take it outside,' and I'm like: "People! We're talking about *a book*!"

As a legal assistant she had some knowledge of the law and asked the board why their attorney was not at the meetings. Bonsell's response—in fact his whole attitude to her—was dismissive, "like 'Go away, little girl.'"

A few days after Beth's letter appeared in the newspaper, school board member Heather Geesey responded.

> This letter is in regard to the comments made by Beth Eveland from York Township in the June 20th York Sunday News. I assure you that the Dover Area School Board is not going against its mission statement. In fact, if you read the statement, it says to educate our students so that they can be contributing members of society.
>
> I do not believe in teaching revisionist history. Our country was founded on Christian beliefs and principles. We are not looking for a book that is teaching students that this is a wrong thing or a right thing. It is just a fact. All we are trying to accomplish with this task is to choose a biology book that teaches the most prevalent theories.
>
> The definition of "theory" is merely a speculative or an ideal circumstance. To present only one theory or to give one option would be directly contradicting our mission statement. You can teach creationism without it being Christianity. It can be presented as a higher power. That is where another part of Dover's mission statement comes into play. That part would be "in partnership with family and community." You as a parent can teach your child your family's ideology.

"With the reelection of George Bush," said Beth, "I think the fundamentalists in the United States felt emboldened, and I think this [case] is a direct product of that, and it frightens me because . . . there were only eleven of us parents. There should have been more."

Although she'd "do it again in a heartbeat," there were some negative consequences. "I did have some Brownie moms who wanted to pull their girls because I was involved, and I'm like, 'You know what, I'm not preaching to your girls, it's all good no matter what they believe, but if you want to pull your girls that's fine, I'm sorry for that, but for me, I'm teaching these girls how to do for themselves, how to do for other people, how to do for your community.' We don't help each other, my God, where will this world be?"

As for her own children, "I want my girls to be accepting of people no matter how they're different, you know, color of your skin, you know, gay, lesbian, who cares, you know, we're all *people*."

When I asked her about her religious faith, she said she was a strong believer in karma. "It's gonna come back to bite ya one way or the other... I've done things when I was younger and, you know, fifteen years go by and something else happens and I think: 'You know, that's what I get—same type of situation but now I'm on the other side of it.' And I'm like 'Damn, that just sucks!' Well you know, that's what I get for doing what I did back then. I just believe that everything you do affects everybody, be it the smallest thing, it affects everybody—*everybody*."

She became more interested in Eastern religions after going to a work-related seminar in Atlanta and hearing Deepak Chopra speak. She now considers herself more or less Buddhist, "although, I'm sorry, I still gotta kill bugs."

A woman with a hot temper, she found that Buddhism and meditation had made her more tranquil. "My mantra is 'Six months from now, how will this affect me?' You know the kids spilling milk at dinner—in the grand scheme of things, how important is this? It's not, it's insignificant, just let it go, it's not worth getting upset over."

When I asked her what her political beliefs were, she laughed and confessed to being "a registered Libertarian. You just want the government out of it, which is part of the problem: we want more Libertarians to get involved—but we're all antigovernment, so nobody *wants* to get involved!"

As I drove away from the house, past the big looming church where Buckingham worships—where every year they put on a passion play ("I mean they have live animals, they're pulling their sheep, their mules")—up along the back roads circling Dover, I began to laugh at the unexpected pleasures of my work and at what's best and worst about America, this: a Libertarian Buddhist Girl Scout leader—plus a lapsed Catholic agnostic husband, two funny kids, and a free-ranging rabbit—all these lively, intelligent creatures (and yes, I mean to include the rabbit)—living opposite a church wherein the worshippers still believe in the literal truth of Genesis.

God bless America, dude!

Pennock qua Pennock, Finches qua Finches

THE SECOND EXPERT witness for the plaintiffs was Robert Pennock, an enthusiastic man with a beard. A professor at Michigan State University, he had a BA, a double major in biology and philosophy, and a PhD in the history and philosophy of science. His primary appointment was in the college of natural sciences, but he was also in the department of philosophy, the college of engineering, the computer science and engineering department, *and* the graduate program in ecology, evolutionary biology, and behavior.

To a high school dropout such as myself, this seemed like a man from whom one might pick up a thing or two.

He first spoke of how evolution was a great exemplar of the scientific method, "a well-confirmed interlinked series of hypotheses," and was useful not only in and of itself but as a way of learning how to think. "One needs to know it with regard to medicine, and even with regard to engineering applications . . . So there's practical applications to evolution right now. You can get a job at Google if you know something about evolution."

We next received a lesson on the history of "methodological naturalism," going back as far as Hippocrates and his attempt to look at various diseases in a scientific way, in particular epilepsy, then known as the sacred disease.

"The idea was that it was kind of divine possession when one went into an epileptic seizure. Hippocrates suggested that we should not think of it in that way but just think of it as a normal illness and try to find a normal, natural way of curing it." He applied the same thinking to epidemics, then thought to be "the result of displeasure of God, perhaps. Hippocrates said we should try to find—by cataloging natural regularities—try to find causes for epidemics."

This was followed by a critique of intelligent design, with particular attention to William Dembski, a big cog in the movement. In an article titled "What Every Theologian Should Know About Creation, Evolution, and Design," Dembski stated: "The view that science must be restricted solely to purposeless naturalistic material processes also has a name. It's called methodological naturalism. So long as methodological naturalism sets the ground rules for how the game of science is to be played, IDT [intelligent design theory] has no chance."

Later in the article, which Pennock read into the record, Dembski wrote, "In the words of Vladimir Lenin, 'What is to be done?' Design theorists aren't at all bashful about answering this question. The ground rules of science have to be changed."

Rothschild, who was examining Pennock, paused a moment and then said:

"And I have to admit I didn't know until I read that, that Vladimir Lenin was part of the intelligent design movement . . ."

Within a few days, Rothschild received an Internet proposal of marriage.

Pennock's cross-examination was by Patrick Gillen.

Gillen began by asking Pennock whether the religious implications of a scientific theory, such as those of the Big Bang, necessarily made it unscientific.

Pennock quickly sliced this up into its constituent parts and disposed of it. One thing had nothing to do with the other. People could believe what they wanted; that was neither his business nor particularly interesting. All that counted was the evidence.

Gillen moved in on a computer program that Pennock and three colleagues had designed to demonstrate evolution, the Ancestor program.

"The Ancestor," Pennock explained, "is simply a self-replicator, an organism that has instructions to allow it to replicate itself but otherwise is just a series of blank instructions." These computer organisms were then dropped into an artificial digital life system designed to test evolutionary hypotheses. The "viruses," if you like, were seen to mutate and develop in this environment. Those that adapted best survived; those that didn't died.

"They evolved things," said Pennock, waving his hands around, "where the programmer would think, 'Why, I would never have even have thought to do it that way!'" Gillen began to ask another question, but Pennock, leaning even further forward in his chair, now bouncing with enthusiasm, was too full of gusto to be stopped.

"And the other thing about it is—sorry, I get excited about this—is that we can keep track of the *full evolutionary history!* So we have a complete fossil record, if you will!"

He beamed at the courtroom, which responded with supportive laughter. Gillen collected himself and pushed on, trying to extract the obvious: all this might be true, but if anyone looked at one of the resulting mutated "organisms," he would actually be correct if he inferred that there was an intelligent designer behind it—four of them in fact.

Pennock would have none of it. Neither he nor Darwin was interested in who created the original organism (a tough concept for Gillen, who clearly had a pretty good idea who He was and had to bite his tongue not to mention Him by name), only in the *mechanism of its development*.

When court finished for the day, I asked Pennock if I could come and see these organisms, hoping there would be some Pac-Man–like creatures to view, but was disappointed to be told that they did not exist in visible form.

I had only one problem with Pennock, and in fact with many of the scientists who spoke: their use of unnecessarily obscure words. As if the science wasn't hard enough to understand to begin with, Pennock would use a word like "qua" instead of "as" or "by virtue of being."

For example, he said, "Sometimes people will speak qua scientist and sometimes they will speak about something from their own personal views." Would he say to his wife, "Listen, honey, this bathroom's a mess, and I'm not just saying this qua husband?" Well, perhaps. But in this context why confuse the antievolutionists, who already have plenty of resentment toward the educated, by using Latin?

If there is a turning away from science in America and elsewhere and a reversion to simpler understandings of the world, this may be part of it. In order to keep hold of its audience, even the church stopped using Latin. There are few endeavors that don't require some element of salesmanship, and maybe science should try to present a more inviting face.

In spite of the above, Pennock remained one of my favorites. His joy in knowledge and his curiosity were palpable. It was the same joy I saw, albeit in rather battered form, in the science teachers at Dover High. They loved science, and they loved trying to convey this love to their students. If only they could be left alone . . .

• • • •

As had been agreed at the previous school board meeting, the science teachers met in private with the curriculum committee—Buckingham, Harkins, and Casey Brown—and Alan Bonsell, the board president.

It was the last day of the 2004 school year, and Bert Spahr, Jen Miller, and Rob Eshbach were dying to get home to start enjoying the summer vacation. It soon became ominously apparent, however, that Buckingham had been reading *Biology* and that their vacations might not be as free from woe as they had hoped.

Buckingham had a list of more than a dozen objections to the content of the book, all of them having to do with evolution. The first of these was a science timeline in the book, which he disliked because it mentioned the publication date of *On the Origin of Species* but did not mention God or creationism. His last objection was to the inclusion of the words "Darwin's finches."

"And what was his objection to the page about Darwin's finches?" asked Rothschild when Carol Brown was on the stand.

"Because the finch had been named for Charles Darwin."

The subject of the mural now came up, with Bert Spahr asking Buckingham how he had obtained the photograph of it and whether his dislike of the picture was motivating his attacks. He did not answer her directly but continued to harp on about the monkey-to-man issue.

When Pat Gillen was cross-examining Spahr, he asked her about Buckingham's concerns with "origins of life." She replied:

"He had asked us more than once if we teach man comes from a monkey. In response to that, in utter frustration I looked at Mr. Buckingham and I said, 'If you say man and monkey one more time in the same sentence, I'm going to scream.' He did not do that, and I didn't have to."

This antisimian bent has always seemed to me both strange and revealing. People who are quite happy to be compared to dogs, horses, mules, and hawks ("He worked like a dog, was as strong as a horse, was as obstinate as a mule, and watched like a hawk") will, when linked to the far more intelligent ape or monkey, become almost apoplectic. Perhaps for some the comparison is just too close for comfort.

Not wishing to digress again, but I'm wondering if perhaps you are saying to yourself: What a snob this man is—all these cracks about these poor hicks and so on. I'd only like to point out that my sympathies are generally with the underdog and that it's usually against my nature to support anything that has even a whiff of orthodoxy to it. It was only when it dawned on me that although the majority of scientists were on the side of evolution, most of the country was not, that I began to feel comfortable with my contempt for the Buckinghams and Bonsells of the world. The worst form of elitism (dominant both in the educated world and in the scientific community particularly) was to so disdain creationists and fundamentalists that you merely sniffed and turned away. They were so inferior that you needed neither to imagine what it was like to be them nor to conceive that they might eventually, through sheer force of will and faith engendered by bitterness or fear, emerge victorious.

To frankly and vocally despise them, as I eventually did, was, in fact, a form of respect.

If it took a while to allow myself to feel this way about Buckingham and Bonsell, it took me a lot less time to arrive at an honest contempt for many of the educated people who encouraged and supported them. This did not include Michael Behe, the main proponent of intelligent design, because I found him to be a likeable, quixotic fellow. The men who ran the Discovery Institute were another matter. For various reasons explained later, I found the whole tank rank.

Out of this rank tank now came another irritation for the teachers. The Discovery Institute had sent Bill a video, *Icons of Evolution*, a seemingly plausible attack on evolution, which he now insisted the teachers watch. This they wearily agreed to do.

"The last thing I said to Mr. Buckingham," said Bert, "was 'Do I have your assurance that we will have the 2002 *Biology* textbook in the hands of our teachers when fall begins?' He looked at me and said yes, and I took him at his word."

Everyone shook hands, and the summer vacation began.

If, as is often quoted, "The only thing necessary for evil to flourish is for good men to do nothing," the summer was an ideal opportunity for the school board to continue its work.

Summertime

WHILE DOWN IN Harrisburg, I always had the Tennessee Scopes monkey trial in mind and would often try to make comparisons.

Could Steve Harvey be compared to Dudley Malone, the Catholic member of Darrow's defense team? Was Eric Rothschild like Clarence Darrow? Vic Walczak was actually closer to Darrow than any of them, but though he and Darrow shared conviction and a willingness to scrap, Darrow was older and more of a showman.

The best comparison I made was between the profiles of Richard Thompson and William Jennings Bryan, which were amusingly similar. I foolishly pointed this out to a reporter—who previously knew nothing of Bryan's appearance—and he used it before I could use it myself.

One of the characters who fascinated me in the Scopes trial was John Scopes himself. As with Tammy Kitzmiller, his name will be forever attached to an important case, but his involvement, like Kitzmiller's, was actually quite minor, almost accidental.

The man who actually taught evolution at the high school in Dayton, Tennessee, was married and did not want to get involved in such a high-profile case. Scopes, who was twenty-four, had been teaching only a year and was not the biology teacher but a general science teacher and football coach. Good-looking, charming, funny, free-spirited, and athletic, he found the daily assemblies so boring that he devised a way to take a group of students down into the science lab, where he let them discuss whatever they liked while he smoked cigarettes and listened.

One day in the summer of 1925, he was playing tennis with some of his students. School had ended four days before, but Scopes, who was from Paducah, Kentucky, had not gone home because of a girl he had met. Having not seen her before, he asked flirtatiously whether this was because she spent her whole life in church, singing hymns and praying. She replied that indeed her family was involved in the church, and if he wanted to see her again he would have to attend a church social during the upcoming vacation.

A small boy arrived and watched the tennis game for a while before going over to Scopes to tell him a group of men up at the drugstore wanted to see him as soon as it was convenient. Scopes finished the set and then, still sweating, strolled through the hot and humid town toward the drugstore.

Dayton had been a coal and iron town until, for complicated reasons, it had fallen on hard times. The men gathered in the drugstore, chief among them George Rappleyea, were looking for a way to put it back on the map. Hosting an evolution versus creationism trial in the wake of the new Tennessee antievolution law might do the trick.

Rappleyea asked Scopes whether it was possible to teach biology without teaching evolution. At that time the drugstore sold the school textbooks, so Scopes, by way of answer, reached up to one of the shelves and took down *Hunter's Civic Biology,* which had been used in Tennessee schools since 1909. Scopes showed them the book's explanation of evolution. When asked if he'd ever used the book, Scopes said he had, although he wasn't sure whether he'd ever taught evolution from it.

This did not bother the men too much. Would Scopes be willing to "let his name be used" in a test of the constitutionality of the new law forbidding such teaching? Scopes said okay.

A couple of attorneys for the city—who disliked evolution but were dubious about the constitutionality of the new law—agreed to prosecute the case. A warrant was sworn out charging Scopes with violating the law, and he went back to his tennis game.

His life was, if not ruined, certainly damaged by the trial. After it was over, he moved to Chicago and took a postgraduate course in geology. Considering a return to teaching at the college level, he applied for a much-needed fellowship. The reply he got was "Your name has been removed from consideration for the fellowship. As far as I'm concerned you can take your atheistic marbles and play elsewhere." Realizing that his notoriety would follow him wherever he went in academic life, Scopes left college and took a job as a geologist with Gulf Oil of South America. He spent several years in Venezuela and Mexico and remained employed in the commercial sector until he died.

All this lay in the future, though. At the time of the trial, Scopes was a carefree man who took the case seriously—but not that seriously. One intolerably hot afternoon, his attorneys suddenly noticed that their defendant was not in court. It was a crucial moment in the trial, but luckily no one else noticed, and after an hour or so, Scopes strolled back in and took his seat. He and William Jennings Bryan's son had gone up into the hills to swim in a waterhole.

I never found my Scopes equivalent in Dover, but when Robert Linker, one of the biology teachers at Dover High, took the stand, I saw some similarities. For reasons I never understood—there was so much about the defense that was inexplicable—Gillen had called him as a hostile witness.

(When I asked one of the plaintiffs' lawyers about a cross-examination of a witness by a Thomas More lawyer—a line of questioning that revealed weaknesses in their own case—he said, "We don't get it either, but the good news is that whatever we forget to bring out during direct, we can rely on them to bring out during cross.")

At the start of the school's summer vacation, Buckingham, perhaps not liking the Discovery Institute's nonconfrontational stratagems, had sought a more combative advocate and found the Thomas More Law Center, "The Sword and Shield for People of Faith." As legal chit-chat is privileged, we never got the full details of his and Richard Thompson's conversation, just two salient facts: that should it become necessary, Thompson would represent the board for nothing, and that Buckingham might want to look at *Of Pandas and People* as an alternative to *Biology*.

Buckingham duly got hold of a copy of *Pandas* and, having "glanced through it," was so excited that by July 25 he had sent a request to Superintendent Nilsen to order 220 copies immediately. Nilsen, in a fit of wisdom, rebuffed him, so Buckingham started pestering his fellow board members and the science teachers, one of whom was Robert Linker.

Linker was a grinning, likeable man who seemed far younger than he must have been, having taught at Dover for well over a decade. Perhaps because of his sporting activities—he was, I think, the wrestling coach—or maybe just because of a Scopes-like disposition, he was less engaged in the dispute than were the other science teachers. As we all knew who he was, but didn't see him until the very end of the trial, I soon dubbed him "the Missing Linker."

Throughout his testimony, no matter what Gillen asked him, Linker smiled and gave his reply in an amiable but not very interested fashion. Gillen now questioned him about his apparent lack of interest in this new issue. Had he, at the end of the 2004 school year, been asked to review the book *Of Pandas and People*?

"No," said Linker, smiling broadly.

Gillen looked at him in a surprised, almost recriminatory way. "You weren't involved in the whole textbook dispute?"

Linker shook his head and grinned at Gillen, and then, as if relishing innumerable good memories, said, "That was *summertime*."

By this point in the trial, all of us had become so consumed by the case—its lies and liars, its theologians and philosophers, its scientific theories and its antiscience theories, its politics, its sociological impact—so obsessively and exhaustingly consumed by it that at one point a local reporter who sat in front of me, a woman, suddenly looked down at her lap, felt for something, and said, "Oh, my God, I've got my pants on back to front," which caused me to laugh so hard I had to lean forward and hold my stomach, and then saw that I was wearing radically unmatched socks—so consumed that Linker's simple response, evoking as it did the image of an easygoing man enjoying his summer while all this ludicrous bullshit whirled around him, brought forth a burst of laughter from the entire court.

"Good answer," said the judge.

. . . .

Good answer, perhaps, but not good news for Bert Spahr and Jen Miller, who now had to slog through *Pandas* to see what it was all about. But then, as they were doing this, and much to their delight, they found a new book, or rather a new *edition* of Miller and Levine's *Biology*, or, as Buckingham put it:

"At first the science department wanted a book that was a 2002 model. We later found out that there was one that came out that was dated 2004."

The teachers, primarily Jen Miller, upon whom much of this irksome duty fell, were excited because this new "model," or so it seemed to Jen, had been . . . toned down. "It seemed to be aware of the controversy, because I thought it took a lot of the more 'controversial' statements out and replaced them with blander language."

When Fred Callahan was on the stand, Muise asked him, "Is it your understanding that that book, the *Biology* book, covers the theory of evolution consistent with its status in the scientific community?"

"No. Actually, I think the message that I gleaned from Bert Spahr was that it was a relatively mild treatment of evolution ... You know, if Darwin's theory is the overarching critical theory that it is, we were making an accommodation to people's religious beliefs by the very selection of that book."

"If Dr. Ken Miller, the author of that book, said that that book represented a theory of evolution consistent with its standing in the scientific community, would you have any reason to doubt that?"

"No. He'd certainly know better than I would," said Fred, "but he's selling books in Texas, too."

I thought Fred's remark was pretty funny but did not fully understand its deeper—and more worrying—context until I read Ken Miller's deposition.

The most recent edition of *Biology* contained the following addition under the heading "Strengths and Weaknesses of Evolutionary Theory."

> Like any scientific theory, evolutionary theory continues to change as new data are gathered and new ways of thinking arise. As we shall see shortly, researchers still debate such important questions as precisely how new species arise and why species become extinct. There is also uncertainty about how life began.

This was inserted, Miller explained, because "as we submitted our textbook to the state of Texas it was clear the only scientific theory ... that any member of the state board of education was interested in seeing strengths and weaknesses for was the theory of evolution."

The curriculum guidelines in the state of Texas were very specific, and in the end Miller and Levine had decided simply to conform to them. This meant the state board of education for the second most populous state in America, with over twenty million citizens, had identical concerns with those of the school board in Dover, population two thousand, and had, in its idiocy, forced an eminent biologist to change a textbook that was now available, in modified form, all over the world.

Jen Miller and Bert Spahr went over the two editions of *Biology* and created a document delineating the differences. In early August there was to be a school board meeting, the agenda of which would now include proposed approval of the new 2004 *Biology* book.

On the weekend preceding this meeting, Carol Brown received a call from the assistant superintendent, Mike Baksa, telling her of Buckingham's big discovery of *Of Pandas and People.* He was now, Baksa told her, going to propose it as an "adjunct alternative" text to *Biology.*

"I came home from work," said Jeff Brown, "and she [Carol] asked me if I would go to the administration building and pick up a copy, because she was livid. She was on the curriculum committee, and Mr. Buckingham was proposing buying a book to add to the curriculum and had not even consulted with her."

When he got to the administration building he learned there were no copies available, but Sheila Harkins had one, which she had finished reading. He could pick it up from her. "So I went to her house. I remember that conversation pretty vividly. The first thing she said to me was 'I think we should buy this book.' I looked at her. I said, 'Sheila, you don't even want to buy the books that we're *supposed* to buy. Why do you want to buy this book that we don't even need?'"

(In his deposition, he said, "We were a year behind on our biology book because of Sheila Harkins, who hates to part with a dime. I can say this with all due respect: Sheila is cheap." Another ex–board member said Sheila was so tight she could "squeeze the nickel 'till the buffalo farts.")

"She said, 'Read the book,'" Jeff continued. According to her, it was "eye-opening" and showed exactly what was wrong with evolution. "I said, 'Fine, I plan to read it; but why are you so in favor of buying this book?' She said, 'Just read the book.'"

All weekend, he and Carol did as instructed, trading it back and forth.

"By the second paragraph," said Jeff, "I felt they were calling me an atheist because I believe in evolution. And that made me furious."

• • • •

On the evening of August 2, thunder rumbled across the skies above Dover as the board members gathered and intoned the Pledge of Allegiance. Jane Cleaver was away in Florida, so only eight board members were present, and the meeting—"summertime"—was sparsely attended. Even Barrie Callahan had taken a vacation.

After a while, the board started to vote on whether or not to approve the purchase of Miller and Levine's new-model *Biology*.

"I am sitting next to Sheila," said Jeff, "and she is voting 'No.' That doesn't surprise me because Sheila never wants to buy any books anyway. I am not shocked. Angie (Yingling) is just as tight ... those two don't shock me." What he was shocked by, because he liked and trusted him (in fact, the two remain friends), was that Buckingham was voting against buying it. He had, or so Jeff thought, promised the teachers that he would approve it.

The first vote was a four-four split, with Bill Buckingham, Sheila Harkins, Heather Geesey, and Angie Yingling voting against buying *Biology* and Alan Bonsell, Noel Wenrich, Jeff Brown, and Carol Brown voting to approve it. Thus, you had one Quaker voting against *Biology* and two creationists, Alan Bonsell and Noel Wenrich, voting for it.

Alan was always a puzzle. I wondered at times whether there was not some other force acting on him, perhaps his father, urging him to make worse and worse decisions. As Jeff Brown said in his deposition, "I was of the opinion that Alan would have been quite happy to teach creationism in school, but he was also aware of the pertinent case law—and Alan is not a bomb-thrower. You will not get that statement out of me about Bill."

As for Noel Wenrich, he, according to Jeff, had been backing away from teaching creationism since Charlotte Buckingham gave her sermon. "After that meeting, Noel began distancing himself more and more from Bill [and] stopped speaking up so much on the subject. He was quite frankly appalled at the publicity we were getting."

Unfortunately, under school board rules a tie vote was a "No" vote, which meant that the purchase of *Biology* was not approved.

"And then Bill drops his bombshell," said Jeff. When Jane Cleaver got back, he would propose buying *Of Pandas and People*. To buy a supplemental text, he needed to get six votes. If he failed in this, he would continue to block the purchase of *Biology*. If it got his six votes, however, then he'd be willing to also buy *Biology*.

"At that point, according to Angie Yingling, I went white. Bill and I began the first of our truly heated exchanges ... I got about as scathingly sarcastic as I am capable of and asked him 'And if we would happen to read this book [*Pandas*] and not be impressed by it, and not think it was worth the taxpayer's money ... ?'

"He said, 'Then you don't get your book. Either I get my book, or you don't get yours.' And I went, 'So noted.' And we continued downhill from there."

Carol pointed out that they were supposed to buy the biology textbook by July 31, and Bill was now proposing putting it off still longer. Sheila said you could buy supplemental texts any time in the year. Carol said yes, but not without a primary text.

"I remember that going back and forth," said Jeff, "and finally after some long and at times heated discussions ... Angie Yingling moved to reconsider the vote, which only a person who has voted 'No' can do."

Afterwards she told Jeff, "I thought you were going to have a heart attack or kill Bill. I didn't know which."

"Don't worry about my heart," said Jeff, "it is in good shape."

Marilyn Monroe Is Alive and Well and Living in Dover

SEVERAL MONTHS AFTER the trial was over, I went to visit Angie Yingling.

She lived in a large, modern, two-story house in one of those high-end developments that seem to have been randomly plunked down in a field, as if the architect had taken the board with his house models on it, magnified it, and lowered it onto the nearest piece of open ground. A sculpture of an angel stood outside Angie's house, its wings spread, to commemorate her husband, who had died a few years back.

It was 2:30 in the afternoon when Angie, who was forty-seven, greeted me at the door with a wide smile, beckoned me in, and offered me a drink, already prepared for my enjoyment.

"It's vodka! Orange-flavored Grey Goose! Grey Goose l'orange!" she said with a pretty good French accent. "I find if you keep the cap off, it lightens it." It was mixed with pink Kool-Aid: "It has vitamin C in it!"

I had come to do an interview, charm cranked up and stratagems prepared, a plumber with his toolkit ready to find the tap that wouldn't open. Instead I found myself sitting in the abundant Jacuzzi of Angie's memories and opinions, which, as I sat passive and almost silent for the next couple of hours, flowed and frothed around me amiably and without cease.

Angie had a mass of bleached blonde hair (her mother was a beautician who lived around the corner), a figure that stretched the definition of voluptuous, and a high, sexy giggle that burst out from the husky tone of her normal speaking voice, rose to a girlish squeak when she wanted to emphasize a point, then sank to a whisper when describing something bad.

She looked the way Marilyn Monroe would have if she had survived until around 1975, and indeed Angie was a huge fan, having read, by her estimation, over a hundred books about her, and owning one hundred and eight pictures of her, many of which hung on walls around the house.

"I do believe that Bobby Kennedy and Peter Lawford murdered her . . . What they did, I think they did state-of-the-art CIA technology to kill her without a trace." They would have done this because Marilyn was "pregnant to Bobby Kennedy, is what I think."

Like Sheila Harkins and Alan Bonsell, Angie, as well as owning a garage, bought property and rented it out or resold it. She wasn't in the same league as Sheila—who went to "a sheriff's sale, a short sale, a tax sale," to buy properties cheap—but she had just bought "a real neat old Victorian" in York.

She owned several expensive cars, at least one of which was bright yellow. "A Corvette, and my truck, and then I have three more in the garage! I don't know how many cars! I like cars; my late husband was an auto mechanic." She led me out to the garage. "You'll love the first one. Oh, I wanted this car forever. 1958 MG! That's a 986 Porsche, and a Cayenne sport utility." About the MG: "I've never even driven it. You have to pull out a button to start it. Oh, it's pristine."

She had a Jack Russell terrier and a German Shepherd named Rockefeller. She was raised Methodist in Dover and was very studious, graduating third in her class with a 98.6 percent average in 1976. She had studied under Bert Spahr, but, unlike everyone else I met, wasn't that fond of her. Bert had married one of her own students back in the seventies, and Angie did not approve.

Some of the board members had been Angie's friends for twenty years. "Our families go back hundreds of years, as immigrants in the 1700s—from England by way of New York . . . now it's to the point where I don't even speak to the Bonsells. I hung on because I thought for sure Alan would change his mind." Alan's wife, she told me, had breast cancer a few years ago and underwent radical surgery. As an adult, Alan was always fairly religious—she remembered him wanting to put religion into the classroom from the start—but around the time of his wife's illness, she thought, he became more religious.

"Would you like some coffee?"

"Just black."

"I'm learning Spanish; that is 'café neeeegro'—Oh, and I made macaroni and cheese, would you like some?" I declined. "Yeah, I'm learning Spanish right now, yeah, 'poco d'Espanol.' I really have it down, like: 'La comida'—I make the boys lunch down at work, you know—'La comida, la buena comida,' and, ah, 'Tu es mi amigo,' you know, 'You are my friend,' 'Donde,' you know, 'Where'—that's handy. There's two of them, one of them is Mexican and French, so he is completely literate in English, the other one, Amando—oh, my, he knows like, I don't know, 'Where's Miguel?' you know, that's 'Donde Miguel,' he knows like that . . . and inspection, 'inspekshionay.'" Angie also spoke French, as evidenced by "l'orange." "Oh, oui, je parle francais."

"I was seventeen when I got married, eighteen when I got pregnant, nineteen when I got divorced. That was one of those practice marriages ... Wanted to be a lawyer, never made it because of that, didn't continue, dropped out of college."

She waited on Bill Buckingham for years at the garage she and her husband owned. "As a friend, as a client? No problem." But when it came to his religion, "Again, increasingly, like Alan, over the years, you can think back, increasingly worse, you know more ... I'd tell them, I mean, like: 'You're pushing religion. You can't do that to people who don't share your religion'—" here she reduced the volume of her voice to a squeaky, emphatic whisper "*—and especially to the kids.* This is paramount, this is the whole key to *everything*.' And then they'll say, 'Oh, it's not religion.' Oh, come on. I guess you can argue that 'till the cows come home, you know, but it seems pretty much like it is to me, that's my interpretation."

The Browns, who were her friends—"I love Jeff and Casey, don't get me wrong"—were "more radical ... I'm not like that, I'm more like 'Hey, let's try to work this out;' I'm not just gonna take my baaaall and go hoooome. Let's talk about it."

The Browns eventually resigned and left her to face Buckingham, Bonsell, and Harkins alone. "Those three: let me tell you, you're up against them; you're shit out of luck—you know what I mean: You are shit out of luck. Oh, yeah ..."

We sat around drinking and laughing for a couple of hours. Her assistant, Roy, dropped by. He was a wiry, skeptically amused atheist who had accompanied Angie to several board meetings to lend support. Then the Browns dropped by, and Angie kept on talking and offering drinks and food and conversation, and it was convivial and really fun. I concluded at the end of it that Angie was, if she could stop her brain long enough to let it think, an intelligent, warm-hearted, but vulnerable woman.

As an outsider, you could easily miss that beneath the conflict, a lot of the Dover combatants still cared for each other. When Angie had to do her deposition, for example, Sheila Harkins, her putative enemy, knew that she was nervous and came along so she wouldn't have to face the thing alone.

. . . .

So, to get back to the meeting, it was not hard to imagine Angie being moved by the students' lack of books. Perhaps she was moved by Joshua Rowand, a student representative on the board, who said he remembered spending only 90 minutes talking about evolution. His co-representative said she couldn't remember talking about it at all.

"We need a biology book," said Rowand. "Is all of this really over a few pages out of a thousand-page book?"

"So she moved for reconsideration," continued Jeff in his deposition. "It then passed five to four, and we bought the books." Sheila immediately slipped Jeff a note "referring to the fact that Angie caved in, because I had this nickname of 'Caveman' for having 'caved in' a long time ago. And she passed this thing that had 'Caveman' with an arrow pointing my way and 'Cavewoman' with an arrow pointing Angie's way ... Bill was furious and came over and dressed Angie down royally ... A short while later, there must have been a recess, because I was on my way out to get a smoke. I needed one desperately ... I heard Sheila really ripping into Angie ... Basically calling her a turncoat and a traitor and that kind of stuff ... Angie was ... she was near tears."

If anyone thought Bill—Marine, police officer, corrections officer—was going to quit at this point, they were very much mistaken.

He already had a new idea.

John Haught and the Teapot of Wisdom

YOU MAY BY now have sensed that I have a slight antipathy to certain aspects of religion, so it may come as a surprise to learn that to me the most beautiful mind in the whole trial belonged to a Catholic theologian, the plaintiff's expert witness, John Haught.

He was a slender, professorial man with glasses, who sat very upright in the witness box and wore at all times a thoughtful, slightly self-deprecating smile. He was the author of thirteen books, among them *God After Darwin*, *Deeper Than Darwin*, and *Responses to 101 Questions on God and Evolution*.

He had a PhD in theology from Catholic University

and had recently retired from chairing the theology department at Georgetown University. He could talk flawlessly for several minutes at a stretch, neither fast nor slow, and with the utmost calm and courtesy.

He began by describing intelligent design as "a reformulation of an old theological argument for the existence of God, an argument that unfolds in the form of a syllogism, the major premise of which is: wherever there is complex design, there has to be some intelligent designer. The minor premise is that nature exhibits complex design. The conclusion, therefore: nature must have an intelligent designer."

Two earlier proponents of this idea were Thomas Aquinas and William Paley.

"Thomas Aquinas was a famous theologian/philosopher who lived in the thirteenth century. And one of his claims to fame is that he formulated what are called the five ways to prove the existence of God, one of which was to argue from the design and complexity and order and pattern in the universe to the existence of an ultimate intelligent designer...

"William Paley, in the late eighteenth and early nineteenth century, is famous for formulating the famous watchmaker argument, according to which, just as you open up a watch and find there intricate design and that should lead you to postulate the existence of a watchmaker, so also the intricate design and pattern in nature should lead one to posit the existence of an intelligent being that's responsible for the existence of design and pattern in nature. And like Aquinas, William Paley also said to the effect that everyone understands this to be the God of biblical theism, the creator God of biblical religion." Intelligent design "simply appeals to more recent findings about the complexity of the world by contemporary science."

In studying the intelligent design movement, he had found that all of the prime movers were "deeply religious people, deeply committed to the cause of the survival of Western theism," and that their "objective seems to me to get at the heart of what they consider to be the source of moral and spiritual decay. And they do this by using a strategic tool, or what they call a 'wedge,' to combat the materialistic worldview which they consider to be inextricably connected to a Darwinian way of looking at life or, more generally, to an evolutionary, biological way of looking at life."

Modern science dates from roughly the end of the sixteenth to the seventeenth century, "and of course the figure that stands out is Galileo. And Galileo is important because he told his accusers, his ecclesiastical accusers, that we should never look for scientific information in any theological source ... From thenceforth to this day, science is a discipline where testability is the criterion of its worth."

In Haught's view, this in no way put it at odds with religion. Science and religion deal with two completely different or distinct realms. They can be related, but first of all they have to be distinguished.

To explain this, he used the following analogy.

"Suppose a teapot is boiling on your stove, and someone comes into the room and says, 'Explain to me why that's boiling.' Well, one explanation would be it's boiling because the water molecules are moving around excitedly and the liquid state is being transformed into gas.

"But at the same time you could just as easily have answered that question by saying, 'It's boiling because my wife turned the gas on.'

"Or you could also answer that same question by saying, 'It's boiling because I want tea.'

"All three answers are right, but they don't conflict with each other because they're working at different levels. Science works at one level of investigation, religion at another. And it would be a mistake to say that the teapot is boiling because I turned the gas on rather than because the molecules are moving around. It would be a mistake to say the teapot is boiling because of molecular movement rather than because I want tea. No, you can have a plurality of levels of explanation ... The problems occur when one assumes that there's only one level.

"And if I could apply this analogy to the present case, it seems to me that the intelligent design proponents are assuming that there's only one authoritative level of inquiry, namely the scientific, which is, of course, a very authoritative way of looking at things. And they're trying to ram their ultimate kind of explanation, intelligent design, into that level of explanation ... And for that reason, scientists justifiably object, because implicitly they're accepting what I'm calling this explanatory pluralism, or layered explanation, where you don't bring in 'I want tea' while you're studying the molecular movement in the kettle."

From Haught's perspective as a theologian, intelligent design was neither a way to prove God nor to expand on him, but quite the opposite. "I think most people will instinctively identify the intelligent designer with the God of theism, but all the theologians that I consider great—people like Karl Barth, Paul Tillich, Langdon Gilkey, Karl Rahner—would see that what's going on in the intelligent design proposal, from a theological point of view, is the attempt to bring the ultimate and the infinite down in a belittling way into the continuum of natural causes as one finite cause among others. And any time, from a theological point of view, you try to have the infinite become squeezed into the category of the finite, that's known as ... idolatry."

Concluding his direct testimony, he said "the God of intelligent design seems to be ... a kind of tinkerer or meddler who makes ad hoc adjustments to the creation, whereas what I would want a child of mine to think of when he or she thinks of God is something much more generous, much more expansive, a God who can make a universe which is, from the start, resourceful enough to unfold from within itself in a natural way all the extravagant beauty and evolutionary diversity that, in fact, has happened. To put it very simply, a God who is able to make a universe that can somehow make itself is much more impressive religiously than a God who has to keep tinkering with the creation."

Richard Thompson, miraculously present for this witness—perhaps because he viewed him as an apostate—began his cross-examination of his fellow Catholic in typically aggressive fashion and soon got slapped down.

"Now, one of the first things you said, Professor Haught, was that intelligent design is an old theory, an old doctrine. Is that true?"

"I didn't put it in exactly those terms. I said it's—"

Thompson interrupted. "What were the terms you used?"

"I said that its foundation in history is the natural theology tradition that's been part of Christianity and Christian thought for centuries."

"Well, we could also trace evolution to antiquity, can we not?"

"Evolution, as a scientific idea, is something that's relatively recent. Evolution as a fact goes back 13.7 billion years."

"I'm talking about people fifteen hundred years ago that were postulating evolution as a means that life could have evolved."

"If it was that long ago, it could not possibly have been a scientific idea. There were ancient philosophers like Heraclitus, for example, who complained that things are constantly in motion. And if you want to call evolution that, then yes, but it's not a scientific idea."

"What about St. Augustine—didn't he postulate that?"

"St. Augustine had the idea that the universe has been seeded with what he called *semines rationales*, rational principles, that over the course of time can unfold very much in the way of the more generous theology that I was talking about at the end of my testimony."

"So merely because you trace a particular idea to antiquity or to old tradition does not in and of itself make that idea invalid, does it?"

"Well, if it's science that you're talking about, then we have to go back to the seventeenth century and look at the methods that science was using and that scientists still use. And that's really what's distinctive about contemporary evolutionary theory, that it employs a scientific method which Augustine did not have."

"Please listen to my question. I didn't talk about scientific theory, I talked about an idea. Now respond to it with reference to an idea rather than a scientific theory."

"One has to be careful," said Haught, as if talking to a child, "of what's called genetic fallacy in logic. That's the fallacy that tries to understand any phenomenon in terms of how it originated. For example, you could say that astronomy originated in astrology and that chemistry originated in alchemy, but you can't evaluate, you can't reduce, the present understanding of chemistry, for example, to what the alchemists were talking about."

Thompson launched into a series of questions about the intelligent design advocates—weren't they just as well qualified, just as intelligent as everyone else? Did their religious views necessarily invalidate their scientific ones, and so on—then on to Michael Behe and his scientific theory of irreducible complexity.

To use an analogy of my own—in no way as beautiful as the teapot analogy, but also dealing with things "working at different levels"—watching these two men was like watching a boxer in conflict with a chess player when the rules of the game favored the latter. Thompson, accustomed to pugnacious combat in the ring of criminal prosecution, lacked the subtle training needed to deal with someone as sophisticated as Haught.

Haught maintained an air of patient amusement, not mocking in any way, just entirely confident.

Thompson quoted from *The Origin of Species.*

"If it could be demonstrated that any complex organism existed which could not possibly have been formed by numerous successive slight modifications, my theory would absolutely break down." Wasn't this the challenge that Michael Behe was addressing?

"That's how Behe considers it, yes."

"And you don't?"

"Well, no, because there are other ways of explaining this so-called irreducibly complex entity, including Darwinian ways."

"Isn't that one of the controversies, though, in science?"

"It's a controversy between Michael Behe and the scientific community."

"So it is a scientific controversy?"

"Well, I pointed out earlier, when I was asked, Do I consider this a controversy? that I don't consider the notion of intelligent design, which is the ultimate explanatory category that Behe appeals to, to be a category within which you can have real controversy, so no, it's not a controversy."

Trying to make a distinction between creationism and intelligent design, Thompson asked: "And there's a difference between 'creationist' and 'creationism,' correct, or is there?"

Haught looked slightly baffled: "Between a creationist—"

"'Creationist' and 'creationism.' Is there a difference in your mind?"

"Well, a creationist is a person; creationism is an idea."

"And creationism is an interpretation of nature which takes the biblical narrative of creation and the sequence of days involved in the creation story corresponding to the Bible literally and factually and then come to conclusions based upon their view of the facts in the creation story. That's pretty compound."

"Yes."

"If you can't understand it, I'll try to repeat it again. Creationism is the interpretation of nature?"

"It's a theological interpretation of nature."

"Which takes the biblical narrative of creation?"

"Narrative or narratives?"

"Narrative."

"Because there are several narratives."

"Well, I'm talking about the Genesis—okay, we'll stay with Genesis."

"Within Genesis there are two creation stories."

Thompson looked momentarily stumped, but pressed on nonetheless.

"And then take that story, or those two stories, however you want to address it, and they take it literally and factually and then come to a conclusion about creation."

"Yes."

"Intelligent design is different than creationism, is it not?"

"Yes, in the same sense that, say, an orange is different from a navel orange."

"Well, I'm going to go back to your deposition, and you were pretty clear that there was a difference, were you not, in your deposition?"

"Yeah, similar to the one that I just analogized."

"You basically, early on—I don't want to test your memory. I'll show you the deposition. But early on one of the first things you said was you disagreed with Barbara Forrest and Pennock as to the way they tied together creationism and intelligent design?"

"Yes, from the point of view of strict logical precision, because not all intelligent design proponents are biblical literalists. I would want to make them distinct from creationists, logically speaking, but as far as the substance of this trial is concerned, there is really no major difference."

"Well, I'm asking the questions not just focused on this trial, but focused on the outside world as to what creationism is and what intelligent design is. Okay?"

"Yes."

"And so there is a difference between creationism and intelligent design, is there not?"

"Yeah, but when you say 'difference,' that's not the same thing as to say 'opposite.'"

"Correct, correct. But there is a difference, is there not?"

"Yes, there's a subtle difference."

"Did you ever say there was a subtle difference before?"

"I don't know. I'm sure I've said it to my students."

"Does intelligent design have to focus on the biblical stories of creationism—of creation, excuse me?"

"Not necessarily."

"But creationism does. Correct?"

"Creationists take the biblical story or stories literally, or attempt to do so."

"Well, on previous occasions prior to this trial, you actually accused Robert Pennock of misleading the public when he conflated creationism with intelligent design theory, did you not?"

"Yes, I said that."

"And what does 'conflated' mean?"

"To confuse or to alloy, to bring together."

"To blend. Right?"

"To fuse or blend."

"To blend?"

Haught sighed. "Yeah."

"Okay. So it is wrong for the court to get an impression that creationism and intelligent design are the same thing?"

"They're not exactly the same thing, but on the issues that really matter, they both, as I said earlier, are trying to bring an ultimate explanation into the category of proximate explanations. So substantively, they are identical as far as what is really important in this particular case."

"Well, you're not the legal expert, are you?"

Unconsciously aping (not wishing to denigrate the ape) Buckingham's similarly insulting question to Bert Spahr, which we had heard a few days earlier, Thompson caused the entire courtroom to go quiet.

Judge Jones, a man of impeccable manners, a man who never left the courtroom without gently touching the bailiff at the door upon the shoulder as if to thank him for his tedious service, could not have been impressed. Haught, whom one could not imagine being rude even to his wife (even given all those frequent demands for tea), smiled a little awkwardly and then quietly replied, "No."

As if to redeem himself with a dash of humor, Thompson said, "I want to talk about genes for a while, g-e-n-e-s." (Genes versus jeans; get it?) He waited for a laugh, which did not come, and then barreled onward. "It's true that Darwinians talk about genes having a mindlike character of survival. Isn't that correct?"

Once again, Thompson was being a literalist. As Haught pointed out, scientists use "that kind of imagery as a popular way of presenting their ideas." No scientists actually think genes have minds.

Soon we were on to the Big Bang, which is of interest to intelligent design advocates for two reasons: one, the Big Bang smacks of instant creation, as if perhaps by God, and two, it was originally proposed by a Belgian priest (Georges Lemaitre, 1894–1966), and was initially rejected, most famously by Albert Einstein, who called the priest a buffoon. ID proponents argue that his idea was rejected because of its religious implications, though in fact there were other reasons as well. Later, this being the way of science, Einstein came to see the merit of the idea, accepted it, and apologized.

The point Thompson et al were always trying to make was this: Well, maybe us intelligent design guys are right too, and maybe the scientific world is rejecting us because of our religious beliefs. During the trial the words "Big Bang" were used so often that they began to be greeted by laughter and sighs.

After a short break we were on to the weak anthropic principle and the strong anthropic principle, these being a pair of related ideas. Given the extraordinary set of circumstances and elements that brought about a universe that enabled man to develop mind, the strong anthropic principle says it must have required some kind of cosmic designer; the weak anthropic principle "simply maintains that obviously the universe was set up for bringing about beings with minds because we're here."

Haught's own opinion was that as he believed in a God who cared that consciousness came about, it was unsurprising that the universe was so constructed, but as he was at pains to point out, this was a theological opinion, not a scientific one.

Slowly descending, we were soon at the old ape-man issue.

"Have we ever found or identified our common ancestor?"

"Not precisely."

"We don't even have an idea who that common ancestor would be, do we?"

"I think we're getting closer and closer, by studying genetics especially, to being able to make more and more reasonable inferences."

"Well, genetics is not going to tell us who the common ancestor is, is it?"

"Genetics is telling us more and more about the story of evolution because as we read the human genome, we can see almost chapter by chapter how evolution came about. Genetics is now one of the strongest—you might say strongest—pieces of evidence for evolutionary science."

"Well, let me give you an analogy. I have some nuts and bolts. I take some nuts and bolts and make a car."

"Yes."

"Okay? That's a car. Then I take some other nuts and bolts and make an airplane. They have the same parts, but does that mean that the airplane came out of the car?"

"No."

"So that if there is a God, that God could use the same kind of genetic material making, you know, a monkey or an ape and making a human being. Isn't that a possibility?"

Soon we descended even further into religious inquisition of the clumsiest kind, inquisition met with, of course, the most refined and well-mannered response. As if Thompson had not heard a thing Haught said about a grander vision of God, the point of this seemed to be to attack his faith as being insufficiently obedient to Catholic dogma. This had nothing to do with the trial, but there was something so illuminating about Haught, and he was so clearly capable of defending himself, that none of the plaintiffs' lawyers raised an objection.

"Now, you also have what I would consider," said Thompson, shaking himself loose as if in preparation for a mighty scrap, "and I'm not a theologian, but I would consider an unusual concept of God. Would you agree with that?"

"What kind of concept?"

"An unusual concept of God."

"No, I thoroughly believe that my understanding of God is completely and thoroughly Christian."

"Do you believe God can be surprised?"

"I don't know."

"Didn't you say that in your deposition, God can be surprised?"

"It's possible."

"Well, if it's possible for you to have said that in a deposition—"

"It's possible that God can be surprised."

"Oh. Does God know everything?"

"Everything that can be known."

"What can't God know?"

"Things that can't be known."

"And what is that?"

"You can't specify it. It's in the region of the unknowable, so therefore the unspecifiable."

"So you put some limits on the ability of God to know everything?"

"No, I don't want to limit God."

"You believe that God started the universe and really doesn't know what's going to happen?"

"If you want me to get into the theology of this, I can. It's very complex, and it requires going back to some chapters in the history of theology where this question was debated between Dominicans and Jesuits to the point where the pope told them both to keep still and stop talking about it. And for that reason—"

"The logic there appeals to me," said Judge Jones.

Everyone laughed. "I'll be very quick, Your Honor," Thompson said, and turned back to Haught. "Do you believe in the virgin birth of Christ?"

"What do you mean by 'the virgin birth of Christ'?"

"The fact that Christ was born from the Virgin Mary."

"You have to put this in context to make this a real question. The stories of virgin births were the ways in which ancient religious communities tried to get across to their followers the specialness of the one who was being born. And so the attempt to be too literal about any of these teachings is, in my view, not to take them seriously. So that question is one that would lead only to a misunderstanding if I were to say yes or no."

"So isn't that a doctrine of the Catholic Church—virgin birth of Christ?"

"It's not in the creed. Well, yes, it is. But it's—there are lots of doctrines in all religions that need to be interpreted in order to be taken seriously."

"Well, that's a pretty serious dogma of the church, is it not?"

"What the church said—if you want to find out what the church said, read Leo the Thirteenth's encyclical, *Providentissimus Deus*, published in 1893, in which he said Catholics should never look for scientific information in the biblical text. So if you're talking about the virgin birth as something that's scientifically true, Catholics, by instruction of Leo the Thirteenth, do not have to go that way."

"And you choose not to go that way?"

"Right."

"What about Adam as the first man?"

"Even the Hebrew Bible uses the notion of Adam in the universal sense for mankind."

Thompson asked the same question about Eve and got the same answer. He was clearly both puzzled and appalled by Haught.

"In your deposition, you talked about the resurrection of Christ, and you indicated that when Christ appeared in the upper room after his resurrection, if we had a video camera going, we would never have captured him."

"Right."

"Captured his image."

"Yes."

"Do you still believe that?"

"I believe this, and so does, for example, Cardinal Avery Dulles, who is one of the most conservative church people around. If you read his book *Apologetics and the Biblical Christ*, he says just that: if people did not have faith, if his disciples did not have faith, they would not have seen anything."

"So it was really a matter of having faith and spiritual vision?"

"No, the faith was evoked by the presence of the sense that Jesus was alive."

"So it was not a fact, a historical fact, that Christ appeared in the upper room?"

"Well, this goes back to what I said about *Providentissimus Deus*. Don't look for simple historical, scientific facticity when there's something much deeper there to look for."

Thompson said, "Thank you," and the inquisition was over.

• • • •

Haught's testimony, apart from being a beautiful explanation of God, was instructive in another way. He and Thompson belonged (though I hate to use the word with Haught, who seemed to belong to no one but himself) to the same church and yet were evidently centuries apart theologically. And here was Thompson, a Catholic, defending Protestants. Twenty years ago a fundamentalist Protestant would have refused his help. Even six years ago when I was in the South, I often heard Protestants fulminating against Catholics, reminding me that they would go to hell. Now, however, Thompson, Catholic or not, was enough of a fundamentalist, and enough of a social conservative when it came to abortion and gay marriage, to qualify as an ally.

Fundamentalists of all kinds have taken the idea of God and whittled it down into an ecumenical baseball bat which all can use to crack the heads of those they fear or hate. In the war against materialism, all allies are welcome.

If Protestants and Catholics have so quickly come to overlook their differences for a greater good, what other alliances lie in the future? One between fundamentalist Muslims and fundamentalist Christians? And why not include fundamentalist Jews? Men like Buckingham, Bonsell, and Thompson have more in common—politically, socially, and spiritually—with fundamentalist mullahs and rabbis than they have with me, or even, as we have seen, with Haught, whose only interest was in the divine.

The Panda Cometh

As carefree Robert Linker enjoyed his summer, his fellow biology teacher Jen Miller continued studying *Of Pandas and People*. Even with her biology degree, she had trouble understanding parts of it and could not imagine it would be comprehensible to her ninth-graders. Looking at the copyright page, she discovered that its latest edition had been published in 1993 and that much of its science was out of date.

At a meeting with the curriculum committee in August, Bert Spahr asked Buckingham whether he understood the book. He clearly did not. School superintendent Dr. Nilsen suggested a compromise: that the book not be handed out to students but simply be made available in the classroom as a reference book. But Buckingham was insistent. He wanted a copy of *Pandas* assigned to every student in the ninth grade to go along with *Biology*. He then left the meeting to go to a doctor's appointment. Alan Bonsell reassured the teachers that not every member of the board was in agreement with Mr. Buckingham.

During cross-examination, Patrick asked Jen Miller, "And again, you left this meeting, as you had prior meetings, thinking that it was generally positive and that some progress had been made, correct?"

Jen Miller replied with a shrug but little conviction. "Sure."

This "Sure" summed up how the teachers felt. They hated every minute of this, they wanted to get back to the already arduous task of teaching science, and they would do whatever was necessary to end this, even if the only way to do so was through one infuriating compromise after another.

They felt beleaguered and alone. But they were not alone. Two organizations on the west coast, the proevolution National Center for Science Education and the antievolution Discovery Institute, had been watching what was going on with interest.

• • • •

Although she had gray hair as well cut as any executive's, and although she dressed in suits befitting the head of an important advocacy group, Eugenie Scott, director of the National Center for Science Education, reminded me of a ten-year-old. It was easy to imagine her hopping about a school science lab, geeky but self-deprecatingly funny in a way that would warn you not to cross her lest she turn that sharp wit and deprecation outward.

Scott's posture was upright, but she bounced from foot to foot as if she needed to look at you from this angle, then that, to gauge whether you were getting what she was saying. Behind wire-rimmed glasses, her eyes seemed faintly mocking, but when you got to know her better you found it was not just you she found amusing; it was life itself.

When the Discovery Institute proudly published a full-page ad listing "100 Scientists Doubting Darwinism" she published a competing list entitled "200 Scientists Named *Steve* Accepting Evolution."

(This later became Steves and Stephanies, whose acceptance of evolution was registered on the NCSE's website in the form of a "Steve-o-meter." At the time of trial the Steve-o-meter had risen to 600. As only 1 percent of Americans are named Steve or Stephanie, each one of them was equivalent to a hundred of the names on the Discovery Institute's more promiscuously gathered list.)

Eugenie had a PhD in biological anthropology from the University of Missouri and was a professor at the University of Kentucky in the early eighties when attempts were made to get creation science taught. Soon scientists and teachers from across the country joined forces to combat it. In their view, creation science was "biblical literalism with a lab coat on," and they were determined to stop it. This coalition eventually evolved into the National Center for Science Education. As its website explains:

> The National Center for Science Education (NCSE) defends the teaching of evolution in public schools. We are a nationally-recognized clearinghouse for information and advice to keep evolution in the science classroom and "scientific creationism" out. NCSE is the only national organization to specialize in this issue.

Eugenie heard about Dover in the spring of 2004, but it was not unique—"So many creationists, so little time," she told me with a sigh—and it was not until she learned that *Of Pandas and People* was being considered that she really paid attention. To Eugenie Scott, the panda depicted on the front cover of the book was not a cuddly bear but a much despised bete noir, and the book was an un*bear*able book, a blatantly religious proposition filled with sloppy scholarship, outdated information, and egregious distortions of fact.

When she heard that *Pandas* was padding around down in Dover, she assigned a staffer to the "flare-up"—the previously mentioned Nick

Matzke. For the next few months, he monitored the situation from a distance, occasionally offering advice to a couple of local citizens who called him. From these conversations and from continuing press reports, it soon became clear that this flare-up was not going to extinguish itself.

• • • •

The other organization that had its eye on Dover was the Discovery Institute, which lay eight hundred miles north of the NCSE up the coast in Seattle. Founded in the early nineties by a trio of oddballs with a peculiar mix of interests, its mission statement reads:

> Discovery Institute's mission is to make a positive vision of the future practical. The Institute discovers and promotes ideas in the common sense tradition of representative government, the free market and individual liberty ... Current projects explore the fields of technology, science and culture, reform of the law, national defense, the environment and the economy, the future of democratic institutions, transportation, religion and public life, government entitlement spending, foreign affairs and cooperation within the bi-national region of "Cascadia."

Its Technology and Democracy Project "supports technology as the key engine for economic growth and seeks to free its natural advancement from the burdens of undue government regulation." Its Bioethics Program "examines a constellation of related issues, including assisted suicide and euthanasia, embryonic stem cell research, human genetic manipulation, human cloning, and the animal rights movement." The Economics division "examines national fiscal and monetary policy and works to foster economic growth by limiting tax and regulatory barriers to businesses and individuals alike."

The program that seemed to deviate from these disagreeable but otherwise somewhat rational aims of the Discovery Institute was the institute's Center for Science and Culture.

Started in 1996, it

> ... supports research by scientists and other scholars challenging various aspects of neo-Darwinian theory; supports research by scientists and other scholars developing the scientific theory known as intelligent design; supports research by scientists and scholars in the social sciences and humanities exploring the impact of scientific materialism on culture; and encourages schools to improve science education by teaching students more fully about the theory of evolution, including the theory's scientific weaknesses as well as its strengths.

The Center for Science and Culture (at one point it was The Center for the *Renewal* of Science and Culture) had more than 40 fellows. Among them were most, if not all, of the leading proponents of intelligent design.

Founded by Bruce Chapman, Stephen Meyer, and George Gilder, the Discovery Institute, was, I soon discovered, originally financed by Howard Ahmanson.

Ahmanson was a lonely rich kid who, when he grew up, was plagued not only with Tourette's syndrome but with arthritis. In search of spiritual answers to this unfair double hit from above (although one might argue that inheriting $300 million or so on one's eighteenth birthday constituted some attempt at compensation), he eventually washed up on the theological shores of a man named R. J. Rushdoony, the Calvinist father of Christian Reconstructionism.

Rushdoony-the-Reconstructionist believed that the world, but particularly the United States, should be governed by biblical law. He wanted to end government-administered social welfare and public schools. Segregation, in his view, was a basic principle, and slavery was okay. Homosexuals should be executed, as should pagans, adulterers, women who got abortions, and, last but not least, disobedient children.

I admire the man. I've never understood how, if you take the bible seriously, you can jump over Deuteronomy, where most of these laws are clearly stated, as if it didn't exist. It *does* exist, it's a *hell* of a good read, and it's one of the most unequivocal segments of the good book. Most of it has to do with stoning people to death.

In a Salon.com article by Max Blumenthal, Ahmanson's wife, who also worshipped with Rushdoony, described him as "quirky in some ways" but pointed out that "to impose the death penalty you need two witnesses. So the number of executions goes down pretty quickly."

Not quickly enough, I would imagine, for any visible participant at a gay parade.

Ahmanson claims to have drifted toward a more moderate point of view since his Rushdoony days, and I suspect his current antimodernist philosophy is probably reflected in a document put out a few years ago by the Discovery Institute's Center for Science and Culture, entitled *The Wedge*.

> The proposition that human beings are created in the image of God is one of the bedrock principles on which Western civilization was built . . . Yet a little over a century ago, this cardinal idea came under wholesale attack by intellectuals drawing on the discoveries of modern science. Debunking the traditional conceptions of both God and man, thinkers such as Charles Darwin, Karl Marx, and Sigmund Freud portrayed humans not as moral and spiritual beings, but as animals or machines who inhabited a universe ruled by purely impersonal forces and whose behavior and very thoughts were dictated by the unbending forces of biology, chemistry, and environment. This materialistic conception of reality eventually infected virtually every area of our culture, from politics and economics to literature and art.
>
> The social consequences of materialism have been devastating . . . we are convinced that in order to defeat materialism, we must cut it off at its source. That source is scientific materialism. This is precisely our strategy. If we view the predominant mate-

> rialistic science as a giant tree, our strategy is intended to function as a "wedge" that, while relatively small, can split the trunk when applied at its weakest points.

It went on to say that the thin end of the wedge was intelligent design. "Design theory promises to reverse the stifling dominance of the materialist worldview, and to replace it with a science consonant with Christian and theistic convictions."

The document lists three phases that would bring this goal to fruition.

The first and second phases, which together were supposed to take five years, involved intelligent design research and promotion. In the third phase, "Cultural Confrontation and Renewal," scientific materialism in general and Darwinian evolution in particular would be vanquished.

This phase would take twenty years. Twenty years to change the rules by which science has operated for centuries? Twenty years to overturn what is considered by 99.9 percent of all scientists and 100 percent of all major scientific societies to be one of the best-supported theories in the history of science? It's hard to believe these people take themselves seriously, but they do, and the Discovery Institute is a surprisingly wily advocate, not least in its strategy of slipping intelligent design into the classroom by "teaching the controversy."

The beauty of the strategy is that it can be applied to anything. If the Discovery Institute encounters another eccentric multimillionaire, perhaps one who wants them to promote the idea that the Holocaust never happened or that the earth is flat, no problem: such theories are just as controversial as intelligent design.

• • • •

Barrie Callahan had been away for the summer but kept in touch by reading the local newspapers. With considerable pleasure she had read that the new edition of *Biology* was to be purchased.

However, when she returned to Dover in September, she learned that Bill Buckingham was now planning to introduce an intelligent design book to accompany *Biology* into the biology class. She bought a copy of *Pandas* and started to read it.

She was appalled. At the first public meeting in the fall of 2004, she asked Alan Bonsell what was going on. If there were plans to buy and use this ridiculous book, as she had read in the newspapers, why wasn't it on the agenda?

Bonsell said the board was still doing research into *Pandas*. It had not made up its mind what to do with it. Buckingham, confirming her suspicions, said that just because the book was not on the agenda didn't mean he was dropping the issue.

Lauri Lebo, who covered the meeting for the *York Daily Record*, asked Bonsell if he was aware that some people were concerned decisions might be made behind closed doors to avoid controversy. "Oh, Lord," he said. "It's a little too late for that."

Callahan remained suspicious, and her suspicions were confirmed when she got the agenda for the first meeting in October. Casually tucked away in an FYI among other business was this remark:

> The superintendent has approved the donation of two classroom sets, 25 each, of "Pandas and People." [The] classroom sets will be used as references and will be made available to all students.

Dodging all obstacles that could not be knocked down, intelligent design now seemed to have arrived at Dover High.

Jeff Brown was not surprised. Buckingham, having resigned himself to not getting the school district to pay for the books, had told him he was soliciting donations. "And at that point, Mrs. Cleaver and Mr. Bonsell both said that he should put them down for a donation."

At the next school board meeting, a member of the public asked the board who had donated the books. He was told that the donation was anonymous. Several other people, both at this meeting and at meetings that followed, asked the same question. At first they were told again that the donor wanted to remain anonymous. Later, when pressed, Bonsell and Buckingham said they did not know who the donor was. The mystery of where the books came from would play a major part in the latter half of the trial and eventually lead to much speculation about perjury charges.

Noel Wenrich and Jane Cleaver announced their resignations from the school board. Noel, the creationist, said, "I was referred to as unpatriotic, and my religious beliefs were questioned. I served in the U.S. Army for eleven years and six on the school board. Seventeen years of my life have been devoted to public service, and my religion is personal. It is between me, God, and my pastor."

The board began to draft a statement to accompany the new books.

One day, the assistant superintendent, Mike Baksa, came to see the science teachers and showed them a draft. It read:

> Students will be made aware of gaps/problems in Darwin's theory and of other theories of evolution, including but not limited to intelligent design.

It went on to say that *Pandas* would be made available to the students as a "reference text."

Bert Spahr, Jen Miller, and Rob Eshbach were okay with pointing out the "gaps/problems" in evolution, but they did not want to mention either intelligent design or *Pandas* because both seemed to be creationism and might invite a lawsuit. A few days later, Superintendent Nilsen came to visit Jen in her room. "I thought it was odd, him sitting down with me—and the first thing he said was 'Jen, we just want to let you know, Mr. Bonsell and I have been talking, and we think you would make a great department head.'" He then went on to tell her that Alan had come up with a potential solution to the teachers' objections, one that would avoid the impression they were teaching creationism. At the end of the paragraph would be added these words:

Origins of life is not taught.

At this point, Bert Spahr arrived unexpectedly. "It caused a little bit of tension with Mrs. Spahr," said Jen, "because he came to see me instead of going to her as department head."

When he'd gone, Bert and Jen wondered aloud what it meant to "make students aware of" intelligent design? How aware? In how much detail? Wasn't "making aware" much the same as "teaching"?

. . . .

The most inflammatory meeting, the one that pushed most people over the edge, the one that really assured that the train wreck predicted by the local press was actually going to occur, was the one that took place on October 18, 2004.

The agenda for the meeting included the following:

"To approve changes to the biology I, grade 9 planned course curriculum guide for the 2004–2005 school year ... copies of the changes have been sent to the district curriculum advisory counsel and the science department."

There was an executive meeting of the board before it went out to face the public. Toward the end of it, school superintendent Nilsen handed out two versions of the proposed curriculum change, one from the teachers, one from the administration. "And we were starting out the door," recalled Jeff Brown, "and Mr. Buckingham said, 'Let's get this thing done. We know what we've got to do. This is taking too long already.' And I looked at him. I said, 'Well, see you on the other side, Bill,' and we went out the door."

No one remembers the order in which people spoke, but there were a lot of them, and everyone remembers that the atmosphere was heated. Barrie Callahan implored the board to hold off from voting until everyone had had a chance to assess the ramifications of what was happening. Bert Spahr said the science department had made every effort to compromise with the board curriculum committee, but "the science department, including all of its members, vehemently oppose the curriculum committee's draft that include the words 'intelligent design' ... It has been deemed unlawful, illegal, and unconstitutional to teach intelligent design." She went on to say the teachers did not teach origins of life but that "the book *Of Pandas and People* has, as its subtitle, *Origin of Life*. This inclusion will open the district and possibly its teachers to lawsuits which we feel will be a blatant misuse of the taxpayers' dollars. We further feel that our many years of professional training and science education have not been considered, and it appears Mr. Buckingham is only concerned with his own personal agenda."

At that point, she turned to Buckingham and asked him, "Mr. Buckingham, are you going to direct my teachers to teach intelligent design if it appears on the written curriculum?" When he did not respond, she said, "If so, that places them in a no-win situation. They now have two choices: to defy the directive of a school board or to go into a classroom and commit what they believe to be an illegal act."

Buckingham was unmoved and proposed the motion that would change the curriculum and ultimately lead to the lawsuit.

It was seconded, but it did not immediately go to a vote.

"We tried," said Carol Brown, "to amend what was being proposed... I think there were something like eighteen amendments."

"It was so confusing," said Jen Miller. "There were so many rounds of voting that I sort of lost track of what they were voting on sometimes... I was sitting at a table with Mrs. Spahr. Dr. Nilsen came over and said, 'Which version do you want?' And we said we want B. And he said, 'Whatever happens, don't clap.' And we had no idea what that meant."

From their point of view, *nothing* that came out of this could possibly be cause for celebration.

When a board member spoke about the possibility of getting sued, Heather Geesey said if that happened all the science teachers should be fired because they, after all, were agreeing with the proposed change.

This was too much for Jen Miller and Rob Eshbach, who both leaped up and went to the podium. Miller told Geesey the teachers did *not* agree with the change, and certainly not the inclusion of the words "intelligent design," which were the words most likely to bring on a lawsuit.

Eventually, when all strategies had failed, the board voted on this language:

> Students will be made aware of gaps/problems in Darwin's theory and of other theories of evolution including, but not limited to, intelligent design. Note: Origins of Life is not taught.

The motion, which also included an addendum mentioning *Pandas* as a reference text, passed, with Bonsell, Buckingham, Harkins, Geesey, Cleaver, and Yingling voting for it. The Browns and Noel Wenrich (still serving on the board until someone was appointed to replace him) voted against it.

Carol and Jeff Brown resigned in protest.

After the meeting Buckingham called Carol an atheist. "I remember her crying at the end and giving me a hug," said Jen Miller. "I remember her saying that she tried." A month or so later when Carol ran into Bonsell, he told her she would be going to hell.

The reactions that reporter Joe Maldonado got from the locals seemed more positive toward intelligent design than those expressed in the meeting.

"Science has evidence about where we came from, as do the Christians," said one parent. "Why shouldn't students be able to talk about both?" Another said she would "like to see the Bible used as a reference book in the classroom ... We took the Bible and prayer out of our classrooms and look at how kids are today. We showed more respect when I was in school."

A member of the high school bible club thought the decision was "awesome." One student just shrugged. "Whatever. School, church, they're both the same thing. I know it's against the law to talk about God in school, but so what if it gets mixed in?"

The last word, though, has to go to Heather Geesey.

"We are not going to be sued," she said. "It's not going to be a problem. I have confidence in the district's lawyers."

Of all the odd and mistaken remarks the woman made—and there were many—this had to be the oddest and most mistaken. The district lawyers had categorically warned the board *not* to do exactly what it had just done.

The Smoking Gun

TAMMY KITZMILLER GAVE her testimony in such a soft voice that the judge asked her to pull the microphone closer to her mouth. Her testimony was brief, too, as if Steve Harvey had decided not to make her endure too much of this.

When he asked her how she had first heard of what was happening on the school board, and she began to say she had read of it in the press, Patrick Gillen stood up to "make sure that we have preserved our standing objection to the hearsay in the newspaper articles."

This was an attempt to exclude the articles by reporters Joe Maldonado and Heidi Bernard-Bubb. Ultimately, after much fighting of many different kinds and for many different motives, both reporters would take the stand and face Ed White (he of anti-abortionist fame).

Gillen's objection was sustained by the judge, and Harvey moved on.

Kitzmiller was a quiet woman, slim and attractive, with blonde hair cut in a shag. Looking younger than her years (late thirties), she was a working mother, shy and self-effacing, in whose eyes you could, however, detect a very real resolution. She might not be enjoying this, but she was not going to be bullied.

At times she seemed distracted, sad even, but it was not until several months after the trial when I went back to Dover and spent an evening with her that I found out why.

From a broken family, her mother of strict Lutheran stock, her father raised in the Church of God, Kitzmiller began to question religion at an early age, initially when she was learning about Noah's ark in Sunday school. "I think even then I was calculating in my mind ... how could Noah and his family repopulate the earth and yet blacks, Indians, Chinese...? You know, it just didn't work for me."

Unlike almost every other person I met down there on either side, she was not a believer and was worried as a plaintiff that this would come out, but it never did. Her close friends knew how she felt but did not speak about it. In this part of the country, atheism was not something to be advertised.

She got good grades in high school, graduated, and went to work as an assistant supervisor at an industrial training center, where she taught handicapped people new skills. She was married at twenty. Coming from a broken home, she acknowledged that she was probably looking for stability.

"And what did you get?" I asked.

She laughed and replied, "Two wonderful daughters and a divorce."

Her first daughter was born when Tammy was twenty-one, her second when she was twenty-three. Her husband was a contractor, and they lived comfortably, but after a while it seemed that weeks went by when they barely exchanged a word. After thirteen years, the marriage ended in divorce.

Kitzmiller was alone for a while and then met someone, and they moved into a pleasant house on a tree-lined street in Dover. During the trial, the relationship was breaking up, and eventually the man moved out. This explained the sadness I had noticed.

An office manager for a landscaping company, she was an intelligent woman, capable and curious. When she first heard about the school board's problems with evolution, she made it her business to understand the issues. When she got to know the lawyers, she was stimulated by their intellectual acuity. Now, getting her older girl ready for college, she sometimes looked back and regretted she had not gone herself.

Living next door to Kitzmiller was a similarly capable and intelligent woman, Cindy Sneath, who also became a plaintiff. In 2004, Kitzmiller had lived next door to Cindy for about a year, but the two women had barely said good morning to each other. Sneath's children were younger and Kitzmiller was absorbed in her daughters and her new relationship. One summer day, Tammy saw Cindy outside in her yard, walked over to the fence, and asked her if she'd read about what was going on at the high school.

Sneath had. She was, if anything, even more angry about the whole thing than Kitzmiller. She had already been on the internet and done some research.

Cindy and her husband owned and ran an appliance repair and installation shop not far from Alan Bonsell's auto repair shop. She was much admired by the plaintiffs' lawyers, one of whom said he'd love to have her working for him as an investigator. She was, he said, the kind of woman who, if you sent her down the street to find something out, would come back with more than you asked for, and all of it would be relevant. She had two children, a boy and a girl aged seven and four. Her educational background was, as she said in court, "Graduated high school, diploma, life lessons, hopefully a dose of common sense." She had no formal training in science but was interested in it because her son was interested in it.

"You know, don't get him started on talking about the NASA space shuttle program."

When she was asked what harm any of this had done to her or her children, she said, "As a parent, you want to be proactive in your child's education . . . obviously I'm not an educator, I have no big degrees. I want to be proactive, but I depend on the school district to provide the fundamentals. And I consider evolution to be a fundamental part of science."

Not long after her first conversation with Kitzmiller, Sneath called Paula Knudsen at the Harrisburg office of the ACLU to express her concern. After the explosive October 18 meeting, more calls like hers began to come in, one of the earliest being from the determined Beth Eveland.

Beth was the first plaintiff to definitively say she would sign on to a lawsuit against the board. Next came Cindy Sneath. Unfortunately, neither of them actually had a child in the ninth grade, which was the one that would be affected by the new policy. Cindy told Paula Knudsen about her neighbor, Tammy Kitzmiller, who did have a kid in the ninth grade and was also unhappy about what was happening.

Paula got excited and said, "Tell her to call me."

Pressure soon began to mount on Kitzmiller to join the suit. She believed it was the right thing to do, but she was a private person and wasn't sure how she would handle the exposure. She spent a long night thinking about it.

"Cindy and I had kind of joked about this in the beginning: you know, this could be big." But now the joke had become a reality.

Two weeks later, on October 29—a few days before Bush would get his second term—Kitzmiller drove into Harrisburg to meet with Rothschild and Harvey in the Harrisburg offices of Pepper Hamilton. After talking to them, she signed on and became the lead plaintiff, whose name would represent all the other parents.

Rothschild and Harvey began to sign up the remaining plaintiffs. Most of them did not know each other, but they had individually concluded that something had to be done. The next time Beth Eveland called the ACLU, they mentioned Kitzmiller's name. "I remember looking for Tammy at a school board meeting, trying to figure out who she was."

Among the earliest plaintiffs to join the suit were Bryan and Christie Rehm. Bryan was already involved in another attempt to change school board policy. Because of the four resignations from the school board, applications were being taken from which the board would make appointments. Bryan applied. By now his position on the matter was well known. At a public meeting at which the applicants were questioned, Buckingham asked him if he had ever been accused of child abuse (or molestation; Bryan is not sure which), and though, obviously, he answered in the negative, he was rejected in favor of a man who home-schooled his kids and said he would never put them in public schools.

With the Browns gone, Bonsell and Buckingham were in control.

A group of Dover citizens soon formed Dover Cares, a slate of candidates who wanted to oust the current school board and throw out intelligent design. Bryan became one of its leading candidates.

• • • •

On November 19, the school district sent out a press release containing the statement the teachers would be required to read in January before students in the ninth-grade biology class studied evolution.

At a board meeting on December 6, Angie Yingling proposed a motion to take another look at the whole matter. When no one would second the motion, she resigned.

"It's like being on the *Titanic*," Yingling told the *York Daily Record*. "Everyone seems to see the iceberg, but no one is steering away."

Just as the war began in earnest, Buckingham went absent. He was not at this meeting, nor at any of the meetings in December. He was neither seen nor heard from for several weeks. Many people, including the Thomas More Law Center lawyers, tried to reach him. Rothschild and Harvey, who would soon want to depose him, were also unable to find him. His wife, Charlotte, did not return calls.

The plaintiffs finally got to meet one another on December 14 at the

Harrisburg offices of Pepper Hamilton an hour or so before a press conference to announce the lawsuit. Beth Eveland had sat next to Bryan Rehm at one meeting, and another plaintiff, Steve Stough, at another, "but we didn't know we were all players, so it all came together at that meeting at Pepper, and it was like 'Ah, okay *now* I know.'"

Every one of the plaintiffs I spoke to—and I spoke to them all—described a sense of pleasure and relief at discovering they were no longer alone.

The suit charged that the school board was trying to establish religion in a public school science class, which was prohibited by the First Amendment to the United States Constitution and "made applicable to the states by the Fourteenth Amendment, as well as the Constitution of the Commonwealth of Pennsylvania."

At the press conference, Vic Walczak said, "Intelligent design is a Trojan Horse for bringing religious creationism back into public school science classes," while Reverend Barry W. Lynn, executive director of Americans United for Separation of Church and State, described intelligent design as having "as much to do with science as reality television has to do with reality."

The *New York Times*, the *Washington Post*, and the *Philadelphia Inquirer*, among many other news outlets, covered the story in depth.

As if the school board had not had enough warnings, the Discovery Institute, the leading proponent of intelligent design, issued a statement distancing itself from what was going on:

"The policy's incoherence raises serious problems from the standpoint of constitutional law."

My own personal theory is that even given their propensity for recruiting oddballs, they'd spoken to Buckingham and deemed this ball too odd to suit their sophisticated game. Representatives from the Discovery Institute nonetheless attended the trial, giving them—and again this is purely my supposition—the option to claim credit if by some miracle things went their way.

Stepping into the bombast-slot left vacant by Buckingham, Thompson grandly proclaimed: "The shot was fired, and it's down range now . . . The board stands fast, and the Thomas More Law Center is ready to represent them . . . We've been ready for many years. We were waiting for a school board that has enough courage to do what should be done."

. . . .

As Christmas approached, time was running out for the science teachers. In January they would have to read the new statement. If they did, they could find themselves on the wrong side of the lawsuit, lumped in with the board, while if they refused, they could be accused of insubordination, and those who were untenured could be fired.

But then, just as the semester was about to end, they got a present from an unexpected source. Mike Baksa came to visit Bert Spahr, bearing with him the donated *Pandas*.

"Oh, my Lord, it was right before Christmas," Bert told me, "and he comes and dumps these three big boxes of books in my closet, and he says to me, 'You're to unpack them, count them, stamp them, and number them.'

"I just looked at him. 'I'm not stamping and numbering anything because I don't want the darned things.' I didn't tell him that, but I was thinking it. So of course I went back to see how many of these darned things we had, and there were three boxes and there were twenty books in each one.

"Now, I was actually taking the packing slip out, because I wanted to know who paid for them. They were donated to the district: *Who paid for them?* But as I'm picking the books out, there in the bottom of the box was this catalog."

The front cover read "Home Science Catalogue: The 10th Anniversary Catalogue—Home Training Tools for Strengthening Home Schools with Practical Science Tools."

"So I get the catalog out . . . and I have it on my desk, and I'm looking through it to see where the book is in the catalog . . . and there's the *Panda* thing sitting there, and I look up at the top—and there in all its glory are the words: 'Creation Science.'

"Oh, my gosh—I ran over to Rob and I said, 'You've got to see this!' And then we took the catalog, we gave it to Jen, who took it home and locked it in a safe. Because we knew at that point we had the smoking gun."

Barbara Forrest and the Panda's Tale

As far as I know, the defense did not object to any of the plaintiffs' expert witnesses except for Dr. Barbara Forrest, a professor of philosophy at Southeastern Louisiana University. They had good reason to want to keep her out. Not only was she everything they would hate as a human being—a secular humanist (an atheist!)—but also she had studied how the creationist movement morphed into the intelligent design movement. She had, in fact, written a book about it: *Creationism's Trojan Horse: The Wedge of Intelligent Design.*

Worse yet, she had recently come into possession of

a smoking gun that was even more incriminating than the one Bert Spahr discovered.

Rothschild began by getting Forrest to give a brief history of creationism and its conflicts with the Constitution.

Before *Kitzmiller v. Dover*, the most recent precedent, and the one upon which much of Judge Jones's ruling would rest, was *Edwards v. Aguillard*. In 1981, the Louisiana legislature had passed a law called "Balanced Treatment for Creation-Science and Evolution-Science in the Public School Instruction." Creation science was the intermediary stage between explicit creationism, based on literal interpretations of Genesis, and intelligent design, which posits a more complex scientific argument and avoids actually naming God.

The father of creation science was an Evangelical hydraulics engineer by the name of Henry M. Morris. Morris theorized that Noah's flood could explain a much younger earth than was suggested by mainstream science: ten thousand years or less as opposed to four billion or more. If this was so, he reasoned, then the theory of evolution would be disproved because evolution required long periods of time to bring about its effects.

The flood theory was one I had run into before. In Tennessee, I went spelunking in a mountain cave with creationist Kurt Wise and a group of young Christians in order to have the theory explained to me in detail.

Flood theorists say there are only two explanations for such things as the Grand Canyon or the discovery of shellfish fossils like the brachiopod a thousand feet above sea level: a little water over a long time or a lot of water over a short time. The Grand Canyon, under the latter theory, was formed by a massive cataclysmic flood, which washed out the gorge in a matter of days; the brachiopod was found in caves such as the one I visited, not because the topography of the land had changed over billions of years, but because the flood had brought them up there.

Henry Morris founded the Creation Research Society and employed Christian scientists to try and bolster this theory. In 1970, he and evangelist Tim LaHaye founded Christian Heritage College, and two years later they created the Institute for Creation Research.

LaHaye, a graduate of Bob Jones University and a pastor with his own ministry, is best known for co-writing the apocalyptic fiction series *Left Behind* and *Left Behind: The Kid Series*, which have sold over sixty million copies. These sales figures are comparable to those of books by best-selling novelists Tom Clancy and John Grisham. One of the books in the series, *The Remnant*, was the best-selling novel in 2001.

In 1974, Morris, in an attempt to frame his arguments in a more scientific fashion, wrote and published a textbook called *Scientific Creationism*.

By the time of *Edwards v. Aguillard*, there had been several court cases involving attempts to bring creation science into the public school classroom. None of them had been favorable to its advocates. The Louisiana law tried to be more subtle than the others, but, although it took a long time, it too was struck down by the Supreme Court on June 19, 1987.

This date played an important role in the development of intelligent design and in the writing and publishing of *Of Pandas and People*.

Of Pandas and People was published by the Foundation for Thought and Ethics, which described itself as "a nonprofit organization founded to promote balance in education through the free expression of fundamental values." Its founder and president was Jon Buell.

Buell was born in 1939, graduated from the University of Miami with a degree in communication arts, and then worked for the Campus Crusade for Christ. After several years, he started the Foundation for Thought and Ethics. In the early eighties, he and Charles B. Thaxton (famous for persuading Korea to include creationism in its schools) decided to produce a biology textbook. They got Percival Davis and Dean Kenyon involved.

Davis was a professor of life sciences at Hillsborough Community College in Tampa, Florida. Kenyon was a professor of biology at San Francisco State University.

Although *Pandas* was not published until 1989, its authors had been working on it for several years beforehand. They had, in fact, started work before *Edwards v. Aguillard* was decided.

In January of 1987, while still hopeful he might find mainstream distribution for *Pandas,* Buell wrote a letter to a prospective textbook publisher: "The enclosed projection showing revenues of over $6.5 million in five years are based upon modest expectations for the market provided the U.S. Supreme Court does not uphold the Louisiana Balanced Treatment acts. If by chance it should uphold it, then you can throw out those projections. The nationwide market would be explosive."

Of Pandas and People went through many drafts with several different titles, a fact that Nick Matzke from the NCSE had learned and passed on to Rothschild et al. The plaintiffs' lawyers issued a subpoena to the Foundation of Thought and Ethics asking them to produce these drafts. After resisting in a court in Texas, and having been ordered to produce them, the Foundation of Thought and Ethics thoughtfully and ethically provided the drafts.

The first was a 1983 table of contents. The second was a later draft now entitled *Biology and Creation* and authored by Dean H. Kenyon and P. William Davis, otherwise known as Percival Davis.

By 1987, the book had become *Biology and Origins*. Later in 1987, it became *Of Pandas and People: The Central Question of Biological Origins*. There were two drafts of this latter title in 1987.

So, in 1987, the year that the *Edwards v. Aguillard* decision came down from the Supreme Court, *three* drafts of the same book were produced.

Rothschild asked Forrest whether she was able to determine the date of each draft.

There were indications, she replied. In one of the 1987 drafts there was no mention of the *Edwards v. Aguillard* decision, which came down in June; in the other two it was mentioned.

In two of the 1987 drafts, there was the following definition:

"Creation means that the various forms of life began abruptly through the agency of an intelligent creator with their distinctive features already intact."

In the *last* draft of the year, however, this definition had been changed. It now read:

"*Intelligent design* means that various forms of life began abruptly through an *intelligent agency* ..." (my italics).

But this was just the beginning. Barbara Forrest had taken the drafts and done a word search. In the first 1987 draft, the word "creation" was used about a hundred and fifty times. In the last 1987 draft, the word was used fewer than fifty times. The word "design" was used about fifty times in the first 1987 draft but well over two hundred fifty times in the last draft.

Similar results were obtained by a word search for "creationism" and "intelligent design." In one of the earlier drafts, the words "intelligent design" was not used at all!

At the start of 1987, *Of Pandas and People* was a creationist book.

After *Edwards v. Aguillard,* it became an intelligent design book.

• • • •

The spiritual father of intelligent design was Phillip Johnson, a retired law professor from San Francisco. Brought up Catholic, he had been born again into the Protestant faith. While on a year-long sabbatical in London, Johnson read the *Edwards v. Aguillard* decision and concluded that so long as the rules of science remained as defined by the American Academy of Sciences—which excluded the supernatural—there was no chance of any similar case being won.

In the book he wrote following this insight, the 1991 *Darwin On Trial,* he rejected this concept as unfair.

> Because creationists cannot perform scientific research to establish the reality of supernatural creation—that being by definition impossible—the Academy described their efforts as aimed primarily at discrediting evolutionary theory.

He went on to say,

> The Academy thus defined "science" in such a way that advocates of supernatural creation may neither argue for their own positions

> nor dispute the claims of the scientific establishment. That may be one way to win an argument, but it is not satisfying to anyone who thinks it's possible that God really did have something to do with creating mankind, or that some of the claims that scientists make under the heading of "evolution" may be false.

Paul Nelson, a fellow of the Discovery Institute, wrote an article about the movement and how it was formed, and Barbara Forrest read some of it into the record.

In June 1993, Johnson invited a group of like-minded people to "a conference at the California beach town of Pajaro Dunes. Present were scientists and philosophers who themselves would later become well known, such as biochemist Michael Behe, author of *Darwin's Black Box* (1996); mathematician and philosopher William Dembski, author of *The Design Inference* (1998), and *Intelligent Design* (1999); and developmental biologist Jonathan Wells, author of *Icons of Evolution* (2000).

Nelson noted that "Of the 14 participants at the Pajaro Dunes conference, only three, microbiologist Siegfried Scherer of the Technical University of Munich, paleontologist Kurt Wise of Bryan College, and me . . . could be seen as traditional creationists."

The idea of the conference was to bring together people of different religious faiths and unite them into a single force to combat evolution and its spawn, scientific materialism. The meeting would lead to both the Wedge Strategy and the Big Tent Strategy.

"The promise of the big tent of ID," Nelson wrote, "is to provide a setting where Christians and others may disagree amicably and fruitfully about how best to understand the natural world as well as Scripture."

In a speech Johnson gave at Coral Ridge Ministries in Florida, he proclaimed, "I have built an intellectual movement in the universities and churches that we call The Wedge, which is devoted to scholarship and writing that furthers this program of questioning the materialistic basis of science."

Johnson's tent, within which his disciples could loiter, was indeed a big one and accommodated many fascinating varieties of believers.

The first disciple (the John the Baptist of the movement?) was Steve Meyer, who met Johnson in London even before the latter wrote *Darwin On Trial*. Meyer obtained a degree in geology in 1980 and went to work for the Atlantic Richfield Company, but he soon became interested in creationism and went back to college to study the history and philosophy of science at Cambridge University, earning his PhD in 1991. Meyer was one of the founding members of the Discovery Institute and, in fact, the man who led his fellow believers to that fountain of cash, Howard Ahmanson.

William Dembski, the only son of a college biology professor who accepted evolution, was brought up Catholic but did not take to religion until he went through a difficult period, probably in his teens. He had a BA in psychology, an MS in statistics, a PhD in mathematics, and a Masters in Divinity from the Princeton Theological Seminary. Dembski, who had trouble finding a university job after graduating, was one of the first people to receive financial help from the Discovery Institute's Center for Science and Culture. He was now a professor of theology and science at the Southern Baptist Theological Seminary in Kentucky.

That these two would gain admittance to the tent was unsurprising. But the tent was also large enough to welcome such men as Jonathan Wells, one of the earliest fellows of the Discovery Institute.

"As Dr. Wells explains it," Forrest told the court, "he had a first PhD in religious studies from Yale. He also attended the Unification Theological Seminary, which is the seminary in the Unification Church of which he's a member, and that church is led by the Reverend Sun Myung Moon ... Reverend Moon urged him to go back to school to get a PhD in biology so that he could, as Dr. Wells puts it in his own words, 'devote my life to destroying Darwinism.'"

Dr. Wells, the family Moonie! Was there ever a tent (revival) this ecumenical?

• • • •

Everyone was wondering how Richard Thompson would mount a defense against this huge body of evidence showing that everyone in the intelligent design movement was not just religious but, by their own numerous statements, motivated by religion from the start.

He didn't. Instead, for the most part, he simply attacked Barbara Forrest.

"When did you become a card-carrying member of the ACLU?" he asked within minutes of opening his cross-examination. This was soon followed by "Are you aware that the ACLU holds that all legal prohibitions on the distribution of obscene material, including child pornography, is unconstitutional?"

"Objection, your honor," said Rothschild. "This has absolutely no relevance to Dr. Forrest's testimony."

"It's as much ... relevant as a lot of the stuff that you put on in this case that had no connection at all with my clients," snapped Thompson.

"First of all, Mr. Thompson," said Judge Jones, "if you're going to argue the objection, you argue it to me, not Mr. Rothschild."

Thompson apologized, but Judge Jones sustained the objection. "A cognizable reason for the question is not a tit for tat."

Thompson elicited the information that Forrest was a "dues-paying member" of Americans United for Separation of Church and State, People for the American Way, and the New Orleans Secular Humanist Association. As if to drive home the ghastliness of the latter, Thompson asked her to read from the organization's statement of principles, among which was this statement:

> We reject efforts to denigrate human intelligence, to seek to explain the world in supernatural terms, and to look outside nature for salvation.

Yikes!

Thompson asked her—redundantly, one would think—whether she believed in the immortality of the soul, but before she could commit herself to her awful reply, she was saved by the bell, or, more accurately, a beep from the court reporter's equipment.

"Hang on," said Judge Jones. "Wendy, are you all right?"

"Objection," said Rothschild.

"Are you objecting to the question or the beep?" asked Jones.

Rothschild laughed, then stated that he objected to Forrest being asked about the immortality of the soul. When the court reporter was asked to read back the question, it was found not to exist. Perhaps even God found it offensive.

Thompson asked Forrest whether she objected to evolution advocates attaching religious or philosophical statements to the theory of evolution. She responded that they had a right to say whatever they wanted.

"And if intelligent design advocates or theorists happen to attach a religious explanation for their theory, would you object to that?"

"That isn't what they're doing. They're not attaching a religious component. Intelligent design is, in essence, a religious belief. It is not a scientific belief with a religious component attached to it."

Thompson asked what methodology she was using that excused Darwinists for their philosophical extrapolations but did not excuse intelligent design theorists for theirs.

"My methodology is to simply make a very careful distinction between people who are not doing the same thing. And that is part of what we call critical analysis, to clarify ideas and to make careful distinctions. That's the methodology I'm using."

"Is there a formula that we can look at?"

"It's part of critical thinking. It's part of recognizing the difference between science and religion. It's part of recognizing the difference between a true statement and a false statement."

"You mentioned critical thinking. And I believe you say you've taught a course on critical thinking?"

"I teach it regularly."

"What is a logical fallacy?"

"A logical fallacy is a mistake in one's reasoning."

"And there are several different concepts under logical fallacy, like lists of logical fallacies, is that correct?"

"There's scores of logical fallacies."

"What is a logical fallacy of ad hominem?"

"The ad hominem fallacy is when you dismiss a person's argument and instead attack a person's character."

"What is the logical fallacy of straw man?"

"Straw man fallacy is when you intentionally misrepresent or weaken a person's argument in an effort to make it easy to refute."

"And what is the fallacy of the genetic fallacy?"

"It is a fallacy of dismissing another person's position based on where it came from, the origin of it."

"So when you attack someone as a creationist, or—excuse me—when you say someone is a creationist, it could very well be a straw man argument, is that correct?"

"Not as I'm doing it, no, sir. Only if I misrepresented a person's position. And I'm not attacking; I am describing. I am simply stating the facts of the case."

The day ended with Thompson suggesting that the plaintiffs' first expert witness, biologist Ken Miller, was a creationist. Although he made Forrest squirm a little in her defense of him, Thompson was as incapable of definitively nailing her as he had been incapable of jailing Kevorkian.

By this point, it was late afternoon and everyone was exhausted. As local humorist Mike Argento wrote in his *York Daily Record* column: "About the time that Richard Thompson started his 3rd year of cross-examination of philosopher Barbara Forrest, it was easy to imagine that at that moment, everyone in the courtroom, including Forrest, who doesn't believe in God, was violating the separation of church and court by appealing to God for it to please, Lord, just stop."

As the judge put it more kindly, "We're probably reaching a point where you can wrap it up for today . . ."

Thompson would wake everyone up in the morning with some of the strangest questions and statements of the entire trial.

• • • •

The bell was rung, and Thompson once again came out swinging.

Had Forrest heard a remark from philosophy professor Peter Singer that "evolution teaches us that we are animals so that sex across the species barrier ceases to be an offense to our status and dignity as human beings."

She had not, but I was not alone in wondering if Peter Singer's 2001 remark (in a book review) was perhaps the genesis of Republican Senator Rick Santorum's jarringly bizarre comment about "man on dog" sex in 2003.

In an interview in his office—Santorum's sanctum sanctorum—an Associated Press reporter asked him why he was opposed to homosexual marriage and got the following reply:

"In every society, the definition of marriage has not ever to my knowledge included homosexuality. That's not to pick on homosexuality. It's not, you know, man on child, man on dog, or whatever the case may be. It is one thing. And when you destroy that you have a dramatic impact on the quality—"

The A.P. reporter interrupted, saying, "I'm sorry, I didn't think I was going to talk about 'man on dog' with a United States senator; it's sort of freaking me out."

"And that's sort of where we are in today's world, unfortunately," Santorum continued blithely. "The idea is that the state doesn't have rights to limit individuals' wants and passions."

Santorum is a religious conservative and favors intelligent design, and one can't help but wonder what's happening inside the brains of such people when they make these odd connections. I'm not for a moment suggesting that Santorum is consumed with canine desires, nor even that if he was he could not stop himself from acting on them, but nonetheless, I have a beautiful dog, more accurately a bitch, and in a discussion about gay marriage her furry little form would not be the second comparison to come to mind.

Thompson had two more quotes to run by Forrest, one by Randy Thornhill and Craig Palmer (respectively a biologist and an anthropologist, and authors of *A Natural History of Rape*) suggesting that rape was a natural biological phenomenon and a product of human evolutionary heritage, and another by Steven Weinberg, winner of the Nobel Prize in physics in 1979, claiming that one thing that motivated his science was a desire to free people from superstition.

What this had to do with teaching intelligent design at Dover High was anybody's guess, but no objections were raised, and Thompson continued on relentlessly.

Exhaustion was setting in as he asked Barbara Forrest why she hadn't counted *all* the words in *Of Pandas and People* and then went on to the story of a scientist who had been persecuted for his intelligent design beliefs, and then back to an interminable examination of other matters already covered, until finally the judge said:

"We've got to move through this witness. We've been on this witness now a day and a half almost."

To quote Argento, "after about a decade and a half," during which "the questioning had gone far afield and sent out such powerful boredom rays that people were falling asleep listening to it in Idaho," it finally ended.

When You Open Your Eyes in Hell

By the time January 2005 came around, a note had been sent out permitting students to opt out of listening to the pre-evolution statement. The teachers refused to read it, so Superintendent Nilsen and his assistant, Michael Baksa, agreed to read it instead. Not wishing to get into any legal trouble themselves, they refused in advance to answer any questions students might have.

"If there were students that objected or parents that opted their children out," Kitzmiller explained, "they left the room, and then an administrator walked in and read the statement, leaving no room for questions or answers, and then they left."

Kitzmiller's daughter walked out. She didn't want to be "singled out, but she also did not feel she needed to be in the classroom if her teacher didn't have to be there."

It was a strange moment. Nilsen and Baksa came into the classroom. The teacher and some students went and stood in the corridor. The statement was read. Nislen and Baksa left. The teacher and the students then walked back in.

Julie Smith was the shyest of all the plaintiffs. She came to testify but was rarely seen again, and I considered myself lucky to get a chance to interview her. Few others had. Recently divorced, she lived with her daughter and two dogs in a pleasant house. Her son was older and in college. Julie had graduated from York College and was a medical technologist. A small, slender woman in glasses, whose blonde hair was tied in a loose ponytail so long it brushed against the rear pockets of her jeans, she spoke almost in a whisper.

When I arrived, she was preparing dinner for her daughter, Katherine, who was in the eleventh grade at Dover High but was not yet home. Smith and I sat in the kitchen, and she told me how she had first heard of the board's actions in June and followed developments through the local papers.

She was a Catholic but had no problems with evolution and found the school board's attack on it disturbing. A lot more disturbing was something that happened later in the same year. While discussing what was going on at the school, Katherine turned to her and said, "Well, mom, evolution is a lie; what kind of Christian are you anyway?"

She then learned for the first time that Katherine, who had several friends from Protestant fundamentalist churches, was a member of a Bible club.

Thompson's cross-examination of her was so aggressive that, given Julie's evident shyness, it made one cringe. When she told how she had gone to her deacon and asked him for his views on the matter, Thompson asked, "Is he a theologian?"

"He's a deacon at St. Rose Catholic church."

"Do you know if he has any particular expertise in Catholic theology?"

Steve Harvey's objection that this was beyond the scope was sustained.

Julie and I continued talking. Having grown up partially in Vermont, she had problems with the redneck aspect of Pennsylvania but found the people friendly. Like many of the other plaintiffs, she had now made several new friends as a result of the trial.

A door banged, and Katherine was home. I expected some version of Julie, a shy child driven to religion by low status on the high school social scale; but this was not at all the case. She was attractive, wore tight jeans, and carried herself with friendly confidence. An honor student, she had just sung in the chorus of a school production of the musical *Oliver Twist*. She played tenor sax and clarinet in the band and was going to Europe with the band in the summer. Looking at her, you might not be surprised if she ended up in a rock band in later life. She even had a tattoo or two.

I asked her how things were in Dover now. She shrugged, glanced at her mother, and said the atmosphere was getting better. Both were clearly happy to have the trial behind them.

I asked her if she was proud of what her mother had done. She hesitated and then said she was glad she'd done it, but she herself would not have wanted to be that much involved.

"And are you more towards creationism or evolution?" I asked her.

"I haven't developed my opinion yet," she offered diplomatically, but when I pushed, she said it was hard to make up your mind when you got one version at school—by which she meant from her friends—and another at home.

I teased her for being so equivocal and tactful when faced with a descendent of Darwin, and she and her mother laughed.

"Smart girl," said Julie proudly. "You don't get too much by her."

"You don't get too much *out* of her, either," I said, and again we all laughed.

She confirmed that a lot of kids carried bibles around, but she did not.

Most of her friends were antiabortion and she was on the fence, but she was totally opposed to the death penalty. She liked Bert Spahr, who had taught her the year before. She was in fact considering a career in science, "genetic counseling," which involved helping people with inherited disorders.

I pointed out that studying genetics could challenge her creationist beliefs, and Julie quickly intervened—as if perhaps she knew this was true and was glad of it but didn't want her daughter deterred from this educational route to rationality by stating it so baldly: Katherine would find out what she really thought in college. "I don't try to tell her what to believe. It wouldn't work anyway; she's very independent."

If I had to hazard a guess, it would be that Katherine will escape the fundamentalism of her friends. It was equally apparent, though, that there had been a time when the relationship between mother and daughter had been very painful, and that the cause of this pain began with men like Behe, was passed down through men like Buckingham, and didn't stop until families were acrimoniously shouting in the kitchens of their homes.

. . . .

After visiting Smith, I went to see Steve Stough, another plaintiff.

Stough was a teacher of life sciences at a York County school and, by the time the trial began had a daughter in Dover's ninth-grade biology class. He was not just a science teacher but also a track and field coach. An estimably well-muscled man, he was a few years younger than me and probably a hundred times stronger. His blond/gray hair was cut into a flattop, and he had the confident but not cocky bearing of an athlete who enjoyed his good health and was proud of the self-discipline required to fight what he nonetheless knew was a losing battle with time.

Adherence to a weight-lifting regimen—evidenced by a massive stack of free weights in the ground floor of the house he shared with his somewhat younger, and equally fit, second wife—was mirrored in an equal adherence to another routine: he read all the local papers cover to cover every day, even when he went on vacation and had to access them online.

"This is my sickness," he said in court, laughing.

He followed events at Dover High from the first day it was reported and read all the letters to the editor and all the editorials on the subject. One of the first people to call the ACLU when the statement was approved by the board, he was particularly irritated by the idea that a teacher could make students "aware" of intelligent design without actually teaching it.

"I've been teaching for twenty-nine years. Everything that I say in that classroom is teaching . . . if I say that one NFL football team is better than another, I'm going to tell you that 80 percent of my kids are going to go back to their parents and say, 'This is what Mr. Stough said, and this is how it is.'"

As the lawsuit developed he was happy to have his newspaper "sickness" put to good use. With help from legal assistant Hedya Aryani, he put together several charts detailing what "effect" intelligent design was having on the local community. He discovered, for example, that of 139 letters to the editor of the *York Daily Record* discussing intelligent design, 86 of them mentioned religion, and that of the 43 editorials in the same paper, 28 discussed religion.

No matter what "purpose" intelligent design had, clearly the "effect" was to stir up religious controversy.

As a reward for becoming a plaintiff and for doing all this hard work, Stough had received a vicious and threatening anonymous letter. It offended him, but I suspect that, being such a strong and confident man, he was not much scared.

This was not the case with Tammy Kitzmiller, who received two even nastier letters.

"When you open your eyes in hell you will think about the time when you went against God," began the first. "The ACLU is trying to wreck our country by getting rid of God and to think that some of the heathen parents are helping them. Does God have to send more tornadoes to wake up our country for trying to get rid of God. How can you be as dam [sic] dumb to think that God did not create this world and the people in it. People like you should get out of the country instead of helping to destroy it. I would be ashamed to walk the street after doing such a wicked thing. Who would ever have thought that the time would come that in America they would fight to get/keep God out of school all because of heathen dummies like you. Start reading Genesis, you will read who created the world. God is not done with destroying our country because of heathens like you, you and your group who live here. God will punish you for doing such a [sic] evil thing. I would hate to be in your shoes. I will be praying that God will make you miserable until you accept him as your Lord and Savior. First you must confess your sins and repent."

The second letter, which appeared to have been written by the same person, was so threatening Kitzmiller called the police. It purportedly came from a woman named Sarah Smith, but there was no address and no such person was found.

"If you speak up you can make a change. You didn't just think of all those schoolchildren that won't hear how God made the world and everything in it because of a dam [sic] dumb fool like you. God says the sins of the parents are passed down to the children. When God decides to strike you or your children down, look out. The ACLU is working all these years to get God out of America and a dumb fool got them to help her. I sure would not want to be in your shoes or your daughters shoes. God hates sin, all these young people being killed in auto wrecks, but when your day comes you think you are smart for what you did, but you are just a damn dumb fool so ignorant to know what God can do to you and your children. I will be watching in the years to come how God punishes you or your daughters. Madelyn Murray [a famous atheist] was found murdered for taking prayer and bible reading out of the schools, so watch out for a bullet."

Kevin Fills the Gaps

Brian Alters, a professor of science education at McGill University and the plaintiffs' education expert, came on and duly testified that indeed evolution was an essential part of a good science education. As what he said, though eloquent and forceful, was often expressed by other experts along with details from their specific fields, I won't elaborate on it. Suffice it to say that it was relatively brief and—at least to me—persuasive.

The final expert witness for the plaintiffs was paleontologist Kevin Padian, who soon revealed that evolution had a problem which intelligent design did not. I.D. had produced no positive research, and, although Michael

Behe later proposed a way in which his idea of "irreducible complexity" could be scientifically tested, no tests had been done, and therefore there was no evidence to sift through. By contrast, evolution, the hammer brought to bear on intelligent design, was festooned with evidence produced by decades of research accumulated through numerous scientific disciplines.

Worse yet, one's capacity to understand evolution was often hindered by the unpronounceable names given to its elements. Nowhere was this more apparent than with paleontology. True, some discoveries had been given names like "Lucy," but more often you were asked to assimilate such words as "Archaeopteryx," "Deinonicus," or "Sinornithosauras."

As if in apology for this, Padian, who was here to refute the "gaps in the fossil record," often said "the names don't matter" or referred to the former inhabitants of these fossilized remains as "critters."

The plum job of conducting the direct examination of the plaintiffs' concluding expert witness was given to Vic Walczak. Both men had graduated from Colgate University, Padian going on to teach seventh-grade life science and biology for a couple of years before attending Yale, where he earned a PhD in biology. His dissertation was on the evolution of flight and locomotion in flying reptiles called pterosaurs, which lived during the age of the dinosaurs.

He was now a professor at the University of California Berkeley in the Department of Integrative Biology, and also the curator in the Museum of Paleontology.

The first gaps he was asked to comment on were those in intelligent design.

One of the central arguments in Behe's "irreducible complexity" idea was that "the strong appearance of design in life is real and not just apparent," but as Padian pointed out, appearances can be deceiving. It appears, for example, as if the earth is flat and the sun goes around it.

Second intelligent design "provides misleading definitions of evolution" and "distorts some commonplace scientific concepts." These he elucidated in great detail before going on to address specific instances in his own field of paleontology.

Pandas claimed that the fossil record was so full of gaps that the case for evolution was undermined by it and the case for "abrupt appearance" supported. Padian said, "I would agree that the fossil record is not complete. It will never be complete. On the other hand, how many intermediates do you need to suggest relationships?"

The first subject of discussion was how fish evolved into vertebrates, animals with backbones, and whether there was any fossil record suggesting that such a change took place. *Pandas* declared that "no such transitional forms have been recovered." Not only was this untrue when the book was written, but later discoveries had made it even less true. And then, five months after the trial, a team led by Neil Shubin of the University of Chicago discovered Tiktaalic roseae, a nine-foot-long fossilized fish in the Canadian Arctic. Tiktaalic was clearly transitional and beautifully illustrated how fins first began to transform into limbs.

Because of this, I won't describe how Padian demonstrated that there were already numerous examples of "links" in this particular evolutionary episode, which *Pandas* simply missed or dishonestly decided to ignore. Suffice it to say that the evidence, even in October 2005, was convincing.

Pandas also disputed the theory that birds evolved from dinosaurs. Here the book's authors could be excused to some extent. There was good evidence for the evolution of birds from dinosaurs when *Pandas* was written, but at that time there were few fossils that revealed anything about the origin of feathers. But in the past decade, "a bunch of remarkable fossils" had been found.

With a series of slides, Padian then explained how discoveries of fossils in northeastern China showed rudimentary feathers developing on clearly nonflying dinosaurs. You could in fact see "black fuzz" at one end of a series of slides he showed and full-blown feathers on the other end, and see how feathers perhaps initially evolved to provide heat, then gradually became instruments of flight.

Far from being created abruptly, as *Pandas* suggested, modern science could now trace the evolution of the bird back to a small carnivorous dinosaur of the middle or late Jurassic period, about 150 million years ago.

After lunch, we were treated to similar lectures on the development of the whale. Because it breathes like a mammal it is thought to have gone in the opposite direction of *Tiktaalic roseae*. In other words, it started on land and then went into the water. The closest living land-based relative of the whale is the hippopotamus. Once again, using slides, Padian showed several fossils that demonstrated how the whale could have slowly left the land, the nostrils moving back to eventually form a single blow-hole, until it finally slid into a fully aquatic life.

Many of the transitional forms supporting this hypothesis had been discovered since *Pandas* was written, but others, such as the Basilosaurids, had been known about since the Civil War. It was possible, however, that the authors of *Pandas* were not being dishonest. Perhaps they were simply ignorant. As Padian explained, "I have never seen them at scientific meetings in my field as far as I know. I've never known them to give papers at those meetings. I've never known them to publish in the peer-reviewed literature of any of the fields related to evolutionary biology or paleontology . . . and I haven't seen their work cited by scientists in those fields . . ."

When asked what he thought of the school board's antievolution policy, he said, "I think it makes people stupid. I think it essentially makes them ignorant. It confuses them unnecessarily about things that are well understood in science, about which there is no controversy, about ideas that have existed since the 1700s, about a broad body of scientific knowledge that's been developed over centuries by people with religious backgrounds and all walks of life, from all countries and all faiths . . . I can do paleontology with people in Morocco, in Zimbabwe, in South Africa, in China, in India, any place around the world . . . We don't all share the same religious faith, we don't share the same philosophical outlook, but one thing is clear and that is when we sit down at the table and do science, we put all that stuff behind us."

Robert Muise's cross-examination of Padian was far briefer than that of Miller, but it was essentially the same. He pointed to paleontological theories that had not worked out, and, overlooking the fact that it was paleontologists who had refuted them, implied that the whole affair was sheer speculation. This being so, why shouldn't the speculations of intelligent design also be included in the scientific canon?

It ended with a protracted whine from Muise about pro–intelligent design scientists being persecuted for their beliefs. As a lot of this was hearsay, and as Padian did not seem to know much about the subject, this line of questioning soon fizzled out.

You Pay Your Nickel and You Go for a Ride

THE LAST PART of direct testimony by all the plaintiffs was their assessment of what harm had been done to them and their children by the board's intelligent design efforts.

Kitzmiller said, "I feel that they have brought a religious idea into the classroom, and I object to that. I do not think that this is good science. There seems to be no controversy within the scientific community, and I would think the biggest thing for me as a parent? My fourteen-year-old daughter had to make the choice whether to stay in the classroom and listen to the statement, be confused, not be able to ask any questions, hear any answer, or she had to be singled out, go out of the classroom, and face the possible ridicule of her friends and classmates."

Steve Stough said that the school board had "usurped my authority to be the one in charge of my daughter's religious education. Intelligent design posits an intelligent designer, which for me they're talking about God. It is a more literal translation of the Bible than I would accept, and I plan on teaching my daughter a nonliteral interpretation."

Barrie Callahan spoke first of practical matters. Her daughter had not had a biology textbook to take home because the school board had delayed its purchase while searching for a text that included creationism. She had received an e-mail from a professor in Texas who warned that he would have a hard time accepting Dover students into his program if they'd been taught that intelligent design was valid science.

The school board was trying to "change the definition of science" by introducing intelligent design as a scientific theory, "but also by demeaning, if you will, the theory of evolution . . . so there's students that will be graduating from Dover not having a clear understanding of what science really is."

Furthermore, introducing intelligent design into biology was tantamount to saying, "okay, it's so complex at this point, it's an intelligent designer. Well, that really stops a student from thinking more about that subject. I mean, I think it's really absurd to think that a school district could hinder a student's natural curiosity into researching an area further."

And then, last, "intelligent design is clearly religious. It's not my religion. I am very upset about the idea of a public school trying to influence my daughter's religious beliefs. And that probably is the most harmful."

Fred Callahan said, "It's a constitutional issue. I'm a taxpayer in Dover. I'm a citizen of Dover. I'm a citizen of this country. I think the heart of my complaint, my wife's complaint, is that this is just thinly veiled religion. There's no question about that in our minds. If you were to substitute where it says 'intelligent design,' the word 'creationism,' which, in my mind, it is, there would be no question that this would be a violation of the First Amendment.

"I've come to accept the fact that we're in the minority view on this. You know, I've read the polls. I think, you know, a lot of people feel that this doesn't cross the line. There are a lot of people that don't care. But I do care. It crosses my line. We've been called atheists, which we're not. I don't think that matters to the Court, but we're not. We're said to be intolerant of other views. Well, what am I supposed to tolerate? A small encroachment on my First Amendment rights? Well, I'm not going to. I think this is clear what these people have done. And it outrages me."

• • • •

The last plaintiff to take the stand was Joel Leib. Everyone awaited his testimony—and the end of this half of the trial—with great excitement.

Joel's family had arrived in Dover before the United States was formed, and most of them still lived there. As Joel put it, "If somebody dropped a bomb on Dover that would be the end of the line." An art teacher at nearby Bradley Academy, he had long thinning gray hair, a beard, and a lanky moustache that drooped down either side of his mouth. When I spoke to him on the last night of the trial, he told me that if the kinds of people who ran the school board ever gained national power, there would be no place in America for people like him. He was in the Jeff Brown mold, a local eccentric, and it was a bold move to put him on last.

He had wanted to appear in court dressed as Darth Vader, but Rothschild had forbidden him to do this. Nonetheless, he was still a wild card.

He was not married but had a significant other, plaintiff Deb Fenimore, with whom he had a son, Ian, in the eighth grade at Dover High. When asked how the change in the biology curriculum had harmed him, he replied: "Two ways. Number one, I've got a child in the school. Intelligent design is not science. Every second that he's either in the class listening to it or out in the hallway objecting to it is a second he's not getting an education and can't be functional in a world market. Let me handle the religious aspect of it."

As to the effect on the community, it was he who stated that the machinations of the board had "driven a wedge where there hasn't been a wedge before. People are afraid to talk to people for fear, and that's happened to me. They're afraid to talk to me because I'm on the wrong side of the fence."

When Gillen, who conducted his cross-examination, pointed out that Ian, when he got to the ninth grade, could opt out of hearing the statement, Joel responded by saying that while this was true, "I teach a post-secondary educational class, and I'm still asking people if they can read or write." This was wasting his son's time, and it was wrong.

Gillen asked him if he thought it was a waste of time because he believed intelligent design was creationism. Leib replied in the affirmative.

"Creationism for me," he explained, "and for probably everybody in this room, is a very personal thing. If you teach it in a comparative religion class, you talk about all religions, not just Christianity, not just Buddhism, not just any particular religion. You look at them, you compare them, you see how they are alike and how they are different. I have no objection to that. I just am telling you it is not science. You're comparing apples and oranges, and there's no place in one for the other. It's like teaching science from the pulpit. There's no place for science from the pulpit."

This was the end of the plaintiffs' case. They had been given every chance to present their arguments; now the other side would get its turn. Or, as the judge put it, with a nod toward Rothschild, "You pay your nickel and you go for a ride. That will have to be it."

PART II

The New Testament OF SCIENCE

The Man from Bethlehem

The first expert witness for the defense, Professor Michael Behe, a Catholic and father of nine, was bearded, vague, and tweed jacketed. He arrived every morning in a flat tweed cap, and every evening when he left he put it back on again. Author of *Darwin's Black Box: The Biochemical Challenge to Evolution*, he was a biochemist and tenured professor of biology at Lehigh University in Bethlehem, Pennsylvania, where he had taught for twenty-three years.

Bethlehem—the birthplace of irreducible complexity!

Behe had a BS degree in chemistry from Drexel University and a PhD in biochemistry, which he received in 1978 from the University of Pennsylvania. He was a member of the American Society of Biochemistry and Molecular Biology and also a member of the Protein Society.

For the past twelve years, he had taught, among several others, a class called Popular Arguments on Evolution, during which evolution and "alternative ideas" were discussed.

He had done experimental research on nucleic acid structure but now was "more interested in theoretical issues rather than experimental ones," among them "the issue of intelligent design in biochemistry and aspects of that." This had been absorbing his attention for the past eight years.

Muise (as previously noted, father of a similar number of children), who was leading him through his direct testimony, extracted the information that he had published nearly forty peer-reviewed articles in mainstream science publications such as *Nature*, the *Journal of Molecular Biology*, and the *Proceedings of the National Academy of Sciences.*

A fellow with the Discovery Institute, Behe modestly explained his involvement as being that "my name gets put on the letterhead, and every now and again, we get together and talk."

Darwin's Black Box had sold over two hundred thousand copies and had been translated into more than ten languages, among them Portuguese, Spanish, Hungarian, Dutch, Korean, Japanese, and Chinese. Behe had written a portion of the second edition of *Pandas* dealing with the blood clotting cascade.

Having established this, Muise asked him whether as a Catholic he found evolution to be inconsistent with his beliefs and whether he had any religious commitment to intelligent design. Behe replied in the negative to both questions. At a certain point, he had just begun to doubt that evolution was as good a scientific explanation as it was deemed to be.

Muise asked him to give a "Reader's Digest" summary of his book.

"Well, in brief, in Darwin's day, the cell was an obscure entity, and people thought it was simple, but the progress of science has shown that it's completely different from those initial expectations, and that, in fact, the cell is chock full of complex molecular machinery, and that aspects of this machinery look to be what we see when we perceive design. They look like they are poorly explained by Darwin's theory. And so I proposed that a better explanation for these aspects of life is, in fact, intelligent design."

Trying to get around the fact that his theories had not been taken seriously by peer-reviewed scientific publications, he pointed to several other scientists who had instead made their cases in popularly published books. Prominent among them was Oxford University biologist Richard Dawkins, an atheist, and author or such books as *The Selfish Gene,* and most recently, *The God Delusion.*

One of the reasons he, Behe, had chosen to write a book rather than publish in scientific journals was that there was such a bias against intelligent design.

Although the argument of irreducible complexity was sold as "brand new," it was, as has been pointed out before, far more "like new." William Paley's argument about someone finding a watch and inferring that there had to be a designer had now been added to by other arguments, among them similar conclusions reached by seeing Mount Rushmore or finding "John Loves Mary" written in the sand.

Behe was soon into his central thesis, which was based primarily on the bacterial flagellum.

The bacterial flagellum is an amazing thing. Without the help of a diagram, it is more or less impossible to describe. It looked like something you might find in a science fiction movie, some strange space vehicle with an odd mix of the highly modern with mechanical features from nineteenth-century industry.

Luckily for the court, Behe had a diagram at which he pointed with a laser pointer. In fact, he pointed at everything with this device. Even when there was only text on the screen—often material he had written himself—a red dot danced inaccurately, and distractingly, across the text. After a while, your head spun from trying to ignore this dot and concentrate on the words.

"The bacterial flagellum," he told us, "is quite literally an outboard motor that bacteria use to swim . . . This part here, which is labeled the filament, is actually the propeller of the bacterial flagellum. The motor is actually a rotary motor. It spins the propeller, which pushes against the liquid in which the bacterium finds itself and, therefore, pushes the bacterium forward through the liquid. The propeller is attached to something called the drive shaft by another part which is called the hook region which acts as a universal joint . . . The drive shaft is attached to the motor itself which uses a flow of acid from the outside of the cell to the inside of the cell to power the turning of the motor, much like, say, water flowing over a dam can turn a turbine . . . It's really much more complex than this. But I think this illustration gets across the point of the purposeful arrangement of parts. Most people who see this and have the function explained to them quickly realized that these parts are ordered for a purpose and, therefore, bespeak design."

Sometimes described as "the most efficient machine in the universe," it looked exactly like a machine, and one could certainly imagine a designer having made it. In fact, as Behe quoted David DeRosier, a professor of biology at Brandeis University, "More so than other motors, the flagellum resembles a machine designed by a human."

There are many extraordinary things in nature—coral reefs, say, or clouds—that do not look like human creations, so it was odd that it was the one which resembled a human invention that made Behe think about God.

Using quotes from both Richard Dawkins and Francis Crick, Behe showed how each admitted to seeing apparent design in nature. Crick, for example, wrote, "Biologists must constantly keep in mind that what they see was not designed but rather evolved."

"So apparently," said Behe, "in the view of Francis Crick, biologists have to make a constant effort to think that things that they studied evolved and were not designed."

The implication was clear: their minds were locked in evolution, so they were unable to open themselves to the possible truth of intelligent design.

Ken Miller had said he had been unable to find any positive evidence produced by the intelligent design movement, only negative attacks on evolution. Behe said he would find it if he'd "look at the structures of the machinery found within the cell without Darwinian spectacles on."

In refuting accusations that intelligent design was not science because it was not falsifiable, Behe quoted from an article of his in *Biology and Philosophy:*

"'In fact, intelligent design is open to direct experimental rebuttal ... In *Darwin's Black Box,* I claimed that the bacterial flagellum was irreducibly complex and so required deliberate intelligent design. The flip side of this claim is that the flagellum can't be produced by natural selection acting on random mutation, or any other unintelligent process. To falsify such a claim, a scientist could go into the laboratory, place a bacterial species lacking a flagellum under some selective pressure, for mobility say, grow it for 10,000 generations, and see if a flagellum, or any equally complex system, was produced. If that happened, my claims would be neatly disproven.'"

Muise asked how long ten thousand generations of growth would take. Would you have to wait for ten thousand years to get results?

"No, not in the case of bacteria. It turns out that the generation time for bacteria is very short. A bacterium can reproduce in twenty minutes. So ten thousand generations is actually, I think, just a couple years. So it's quite doable."

Rather than ask him why, if it was so "doable," he hadn't done it, Muise asked whether he thought evolution was falsifiable. Once again, Behe quoted an article of his.

"'What experimental evidence could possibly be found that would falsify the contention that complex molecular machines evolved by a Darwinian mechanism? I can think of none.' So again, the point is that, I think the situation is exactly opposite of what many arguments assume, that ironically intelligent design is open to falsification, but Darwinian claims are much more resistant to falsification."

• • • •

This ended his first morning of testimony, and all the reporters hurried outside to bombard him with questions, I among them.

Over the next few days I often encountered Behe on the courthouse steps. He was a likeable man, and when he found out I was a Darwin descendant, he was delighted, stating later in a newspaper article that I was a friendly fellow and my presence in the courtroom was a comfort to him.

I had two problems with his theories. First, if an intelligent designer had made the bacterial flagellum, it seemed logical to assume he had made everything else, and if so, wasn't this by definition God? One day I was having this debate with him when another man weighed in on my side, suggesting that if you were to infer design from the complexity of the bacterial flagellum, and if you knew that complex machines like the space shuttle were designed by a team, wasn't it also logical to infer that the bacterial flagellum was made by a *team* of gods?

Behe smiled tolerantly—albeit with a touch of irritation—and shrugged: that was not his problem, he himself believed in a single designer. This was his personal opinion (and, he pointed out, it was also Miller's opinion), but this had nothing to do with his scientific claims. As for us, we could believe what we liked.

I then launched into my second question.

If you went back in time, couldn't you point to any number of phenomena that, because we did not understand them, seemed possible only through the intervention of God? One by one, each turned out to have natural explanations that even Behe accepted, so wasn't this likely to happen with whatever mystified us now?

I missed the point, he told me: the bacterial flagellum was not only complex but *irreducibly* complex. In other words, if you removed one part of it, none of the others had a function, so the whole could not have developed by natural selection but must have been abruptly created.

. . . .

In the afternoon, Behe disputed the National Academy of Sciences' definition of "theory," in particular the implication that because something is a theory it must be right. Many previously accepted theories had been proved wrong. He believed that Darwin's theory of evolution would soon join them.

He conceded that "evolution as such, common descent, multiplication of species, those are well tested. The claim of gradualism is in my opinion mixed." (Many prominent evolutionary scientists now agree that some evolution may occur relatively quickly.) "But the component of Darwin's theory, natural selection, the mechanism Darwin proposed for evolution, is very poorly tested and has very little evidence to back it up ... The only focus of intelligent design is in the mechanism of evolution, or the question of whether or not aspects of life show the marks of intelligent design."

There followed an extensive list of things that evolution, in particular natural selection, could not—according to Behe—explain. Among these were the transcription of DNA, new protein interactions, and, of course, his favorites: the bacterial flagellum, blood clotting, and the immune system.

Things looked pretty bad for Darwin. As Behe said, "If Darwinian theory is so fruitless at explaining the very foundation of life, the cell, then that makes a person reasonably doubt whether some other explanation might be more fruitful."

Darwin was even a failure when it came to sex. Under Darwinian logic, "one goal of an organism ... in terms of better evolutionary result, is to get more of the organism's genes into the next generation." As organisms that proliferate by clonal reproduction (without screwing, that is) pass on *all* their genes, they should have an evolutionary advantage. And if this was so, most organisms ought to be doing it that way.

This made no sense to me. How could stasis bring about improvement? Wasn't this what random mutation was about? Wasn't failure part of success? Who cared what the organism's goal was as an individual matter? Surely it was the individual goal versus environment plus time that brought about change and improvement. And, besides, sex is fun, and doesn't being happy give one an advantage in the struggle for existence?

Last, putting aside the fact that asexual reproduction would solve many problems Catholics have with premarital sex and abortion, who'd want to be a worm?

Clearly not the Catholics, because one thing was certain: the three lead attorneys for the defense were doing a far better job of passing on their genes than the three lead attorneys for the plaintiffs. Thompson, Gillen, and Muise had, between them, produced twenty offspring, and Muise and Gillen were still young enough to sire another ten apiece. Meanwhile, Rothschild, Harvey, and Walczak had only managed to produce a scant seven!

Laugh if you like—briefly—but this is actually pretty scary. Fundamentalists are outbreeding other groups, and if the apples don't fall too far from the trees, we're soon going to be overrun by them.

Behe's next big attack on Darwin was the failure of his theory to explain the origin of life. If some spark was required at the beginning, something "extra" to kick-start the whole process, why would that "extra something" not continue to chip in now and then?

In order to show how amazingly complex irreducibly complex organisms could be, Behe spent the rest of the day expounding on various examples in such detail that it would be impossible to write them here without sending you to sleep. There was a lecture on hemoglobin, the Type 3 secretory system, the DNA sequence of whales and hippos, and the lac operon and its relationship to *Escherichia coli*.

Toward the end of the day, Muise said, "Your Honor, we're about to move into the blood clotting system, which is really complex."

"Really?" said Judge Jones, ironically. "We've certainly absorbed a lot, haven't we?" He asked if Behe was available to continue the following day.

"He's available, Your Honor, for as long as we need him," Muise replied.

Judge Jones turned to Eric Rothschild. "Any objections if we . . . ?"

"No," said Rothschild. "He started it."

This was a warning: the more Behe and Muise put up, the longer it would take Rothschild to knock it down, but he would do so.

. . . .

The next morning, the blood clotting cascade was described in blood-curdling detail, as was the immune system, to which most court observers soon became immune, only to rejoin the case when it became comprehensibly philosophical again.

Ken Miller had written, "God made the world today contingent upon the events of the past. He made our choices matter, our actions genuine, our lives important. In the final analysis, He used evolution as the tool to set us free."

Behe did not like this at all. "So here is a scientific theory which is being used to support the idea that we are free, we are free, in apparently some metaphysical sense, because of the work of Darwin."

The oft-quoted line of Dawkins, "Darwin made it possible to be an intellectually fulfilled atheist" came up, as did Darwinian philosopher Daniel Dennett's approving description of Darwinism as a universal acid that destroys our most cherished beliefs.

Larry Arnhart, a professor of political science at Northern Illinois University, wrote, "Darwinian biology sustains conservative social thought by showing how the human capacity for spontaneous order arises from social instincts and a moral sense shaped by natural selection in human evolutionary history," and Peter Singer, a Princeton philosopher, believed that we should incorporate a Darwinian ethic of cooperation into our political thought.

"So the gist of Professor Singer's book," sneered Behe, "is that Darwinian ideas support a liberal political outlook. And he argues for that. So, again, all of these people see profound implications for Darwin's theory far beyond biology."

Of course, even within Behe's world, there was both philosophical and scientific disagreement, so the next order of business was to make a distinction:

Intelligent Design Is Not Creationism.

"Creationism is a theological concept," stated Behe, "but intelligent design is a scientific theory which relies exclusively on the observable, physical, empirical evidence of nature plus logical inferences. It is a scientific idea."

It was a scientific idea, however, that was constrained by science's current definition of itself. Behe believed that any constraint on what conclusions science could reach "hobbles all of science. Science should be an open, no-holds-barred struggle to obtain the truth about nature. When you start putting constraints on science, science suffers."

"Despite these constraints," Muise asked, "does intelligent design still fit within the framework of methodological naturalism?"

"Yes. Despite the constraints, it certainly does, just as the Big Bang theory does."

After a brief foray into directed panspermia and the search for extraterrestrial life, Behe concluded his direct testimony by saying that in spite of some quibbles he had with *Pandas,* he still supported its use at Dover High because it was "very useful for a student to view data from a number of different perspectives."

Muise asked his last—and, given the above lecture, redundant—question: "Does Dover's policy at issue in this case support good science pedagogy?"

"Yes, I think so."

And with that, the sturdy procreative Muise turned over his fellow highly reproductive Catholic to the more sperm-conservative Jew, Rothschild.

• • • •

If Ken Miller was the star scientific witness for the plaintiffs and therefore went first, Behe was clearly the star scientist for the defense and had opened its case.

Rothschild's manner in court was often humorous, invariably smart, but rarely arrogant. He reminded me of a precocious, slightly self-deprecating boy who had been given the chance to take on the school bullies in a debate in front of the headmaster. The rules were on his side now, but he wasn't going to say anything too inflammatory in case he turned the headmaster against him. If the trial was to be won or lost on science, however, this was the pivotal moment. Rothschild and his side had laid all their scientific evidence on one end of the seesaw; Muise had extracted from Behe everything that could reasonably be extracted from him and placed it on the other. It was now Rothschild's job to walk back out onto his end of the plank and start jumping.

Everyone was aware of how significant and potentially dramatic the next few hours would be, and the court was jammed. It reminded me of a similar moment in the Scopes trial when the lawyers for Scopes were pleading to have their all-important scientific witnesses included. William Jennings Bryan, who had stated to the press that this was to be a "duel to the death" between evolution and revealed religion, spoke for the first time in the trial to argue for the exclusion of those witnesses. He did so in a manner very much in accord with the notion of a duel, albeit a mocking, bombastic one. He ended with these words:

"The Bible is not going to be driven out of this court by experts who come hundreds of miles to testify that they can reconcile evolution—with its ancestors in the jungle—with man made by God in His image . . . The facts are simple, the case is plain, and if these gentlemen want to enter upon a larger field of educational work on the subject of evolution, let us get through with this case and then convene a mock court, for it will deserve the title of mock court if its purpose is to banish from the hearts of the people the Word of God as revealed."

He received tremendous applause and sat down well satisfied.

It then fell to Dudley Malone, one of Scopes's lawyers, to argue against him. Malone had at one time worked for Bryan when he was running for president, which he did three times, and so, apart from the fact that his argument was vitally important to the defense, there was an underlying human drama.

Malone knew this. The lone Catholic on the defense, he was exceedingly well dressed and fastidious and was the only attorney on either side who had never taken off his jacket in spite of the incredible heat. This fact had been widely noted in the press. Malone waited until the applause came to an end. He then got up from his chair and stood in the body of the courtroom, thinking. After a few moments, he very slowly removed his jacket, folded it carefully, and laid it over the back of a chair.

The court became silent. Malone began to speak in a soft, reasonable, almost hurt tone of voice.

"My old chief ... I never saw him back away from a great issue before ... We have come in here ready for battle. We have come in here for a duel ... but does the opposition mean by a duel that our defendant shall be strapped to a board and that they alone shall carry the sword? Is our only weapon—the witnesses who shall testify to the accuracy of our theory—is our only weapon to be taken from us so that the duel will be entirely one-sided? That isn't my idea of a duel."

There were murmurs of agreement as he continued with increasing passion. Finally, he said, his voice much louder now: "There is never a duel with the truth. The truth always wins and we are not afraid of it. The truth is no coward. The truth does not need the law. The truth does not need the forces of government. The truth does not need Mr. Bryan. The truth is imperishable, eternal, and immortal and needs no human agency to support it ... We are ready. We feel we stand with progress. We feel we stand with science. We feel we stand with intelligence. We feel we stand with fundamental freedom in America. We are not afraid. Where is the fear? We meet it. Where is the fear? We defy it. We ask your honor to admit the evidence as a matter of correct law, as a matter of sound procedure, and as a matter of justice to the defense in this case."

The applause was so thunderous—and this from people who were philosophically opposed to evolution—that the judge had to hammer his gavel for several minutes before it died down. A policeman repeatedly slammed his nightstick on a table so hard he split it. When another officer came up to help him restore order, he said, "I'm not trying to restore order! Hell, I'm cheering!"

Although Malone lost this battle, and although he and Darrow lost the entire trial in a strict legal sense, this was the turning point in the court of public opinion. Court adjourned and reporters rushed off to file their stories, many of which contained Malone's speech in its entirety.

Scopes stayed behind, sitting next to Malone, as the courtroom gradually emptied out. Soon only he, Malone, and Bryan remained.

Scopes reports that the "Great Commoner," the author of the marvelous "Cross of Gold" speech, the most brilliant speaker of his generation and a man who, before he became a fundamentalist buffoon, proposed and supported some of the most enlightened legislation of his time, sat in his rocking chair over by the prosecution table, fanning himself with a palm leaf fan.

Every now and then he would let the fan drop and just stare ahead vacantly. Eventually, without turning to look at Malone, he spoke.

"Dudley, that was the greatest speech I ever heard."

"Thank you, Mr. Bryan," Malone replied softly. "I am sorry it was I who had to make it."

. . . .

Well, it wasn't quite so dramatic the day Rothschild started his cross examination of Behe, but he was certainly aware of the suspense in court, and there was, as with Malone's removal and folding of his jacket, something theatrical about the slow, deliberate way he prepared for this all-important cross-examination.

He came to the lectern with many documents and carefully arranged them. He then slowly poured himself a glass of water, had a sip, and then looked toward Behe, who by now was adjusting his tie, fidgeting with the mike, and blinking more often than he had before.

"Good afternoon," said Rothschild.

"Good afternoon, Mr. Rothschild."

"How are you?"

"Fine, thanks."

Did Professor Behe have all the documents he might need, asked Rothschild: his deposition, his expert witness report, and so on? Behe replied that he did.

"And I saw," said Rothschild, "that you had a copy of *Pandas*, but do you have a copy of *Darwin's Black Box* with you?"

"No, I don't."

"I am surprised you're ever without one," said Rothschild, smiling as he gave him his own dog-eared paperback.

And then he began.

Putting a paragraph from *Pandas* up on the screen—"Intelligent design means that various forms of life began abruptly through an intelligent agency"—Rothschild pointed out that it did not say life *appeared* abruptly but said that life *began* abruptly.

"Well, that's certainly the word it used, but we can ask, how do we know it began abruptly? The only way that we know it began abruptly is through the fossil record."

"But 'beginning' is different than 'appearances' in the fossil record, correct, Professor Behe?"

"I don't take it to mean that way, no."

"Now, you said you wouldn't have described intelligent design this way, correct?"

"Yes."

"But that's how it's being described to the students at Dover who go to look at the *Pandas* textbook."

"Well, that's one of the places, yes."

"And would you agree with me that if one substituted the word 'creation' for 'intelligent design' there, 'Creation means that various forms of life began abruptly through an intelligent agency with their distinctive features already intact: Fish with fins and scales, birds with feathers, beaks and wings,' the statement would be equally apt?"

"Well, I think the sentence as it is drafted is somewhat problematic, as I said in my direct testimony, so I would not say that either one was apt."

"That's not a good definition of creation or creationism?"

"I don't think so, no."

"Would it be a good definition of special creation, Professor Behe?"

"I don't think so either."

Rothschild asked him about his definition of "theory" and how it applied to intelligent design. Behe agreed that his definition did not match up with the National Academy of Sciences' definition, but he argued it was still valid. It was just a "broader" definition and one that would include intelligent design.

Under his definition, Rothschild suggested, wasn't astrology also a scientific theory?

Behe admitted that it was because under his definition of the word a theory didn't have to be correct. It was just "a proposition based on physical evidence to explain some facts by logical inferences. There have been many theories throughout the history of science which looked good at the time which further progress has shown to be incorrect. Nonetheless, we can't go back and say that because they were incorrect they were not theories."

"Has there ever been a time when astrology has been accepted as a correct or valid scientific theory?" Rothschild asked.

Behe admitted that "the educated community has not accepted astrology as science," but at one time it did. "Simply because an idea is old, and simply because in our time we see it to be foolish, does not mean when it was discussed as a live possibility, that it was actually not a real scientific theory."

Using Behe's own statements about the number of seminars and speeches he had attended, Rothschild suggested that his ideas *had* had a fair hearing but were still not accepted by a single major scientific organization. In fact, both the National Academy of Sciences and the American Association of Scientists had put out statements explicitly rejecting them.

Behe shrugged off both statements. They were "political" rather than scientific. "There are no citations here. There's no marshaling of evidence. As I've tried to show in my testimony yesterday and today, if you actually look at these things, we have marshaled evidence; we have proposed means by which our claims can be tested."

In spite of that, Rothschild said, "in the ten or fifteen or twenty years—or if we go to Paley more than two hundred years—intelligent design has failed to make its case to the scientific community, correct?"

Behe said, "If you look at such statements in booklets issued by the National Academy, they are certainly very hostile to the idea of intelligent design. But it's been my experience that a number of people are interested in the idea. Nonetheless, it's the nature of bureaucracy, I think, to issue statements like this. So I do not consider these representative of the scientific community."

But in fact, it was not *just* major scientific organizations that had opposed his ideas. Wasn't it true, Rothschild asked that "your own university department has taken a position about intelligent design?"

"Yes, they certainly have."

Rothschild asked to have his next exhibit put on screen, a statement issued by the Lehigh Department of Biological Sciences, which he then read out loud.

"'The department faculty . . . are unequivocal in their support of evolutionary theory, that has its roots in the seminal work of Charles Darwin and has been supported by findings accumulated over 140 years. The sole dissenter from this position, Professor Michael Behe, is a well-known proponent of intelligent design. While we respect Professor Behe's right to express his views, they are his alone and are in no way endorsed by the department. It is our collective position that intelligent design has no basis in science, has not been tested experimentally, and should not be regarded as scientific.'

"So you've not even been able to convince your colleagues, any of them, Professor Behe?"

By this time I had been watching Michael Behe for a day and a half. He had sat forward in his chair, earnest and concentrated; but as Rothschild read the statement his posture changed completely. He put his hands behind his head and leaned back in his chair, smiling defiantly. It was such a blatant and out-of-character display of casual indifference, even amusement, that I could not help thinking that this rejection by his own community of scholars was intensely painful to him.

. . . .

In his direct testimony Behe conceded that "evolution as such, common descent, multiplication of species, those are well tested." *Pandas,* however, stated, "The fossil record shows that most organisms remain essentially unchanged. The conclusion to be drawn is that major groups of plants and animals have co-existed on the earth independent of each other in their origins."

This was clearly a strong statement *against* common descent, which proposed that all life evolved like the branches of a tree from a *single entity* billions of years ago.

As Behe was a contributor to the book, and here to defend its use in the high school, he had to do his best to do just that. He wriggled around the problem for a long, long while, but finally and rather irritably he admitted that "*Pandas* is making a negative argument against common descent to . . . enhance the plausibility of the alternative of intelligent design, that's correct."

"Thank you," said Rothschild.

Behe insisted on restating his position. Natural selection was "real, and certainly explains a lot of things," but it was unable to explain other things.

What about intelligent design, asked Rothschild a little sardonically. What did it explain? Wouldn't it be accurate to say that it did not describe who the designer is or even how the design occurred?

"That's correct, just like the Big Bang theory does not describe what caused the Big Bang."

Intelligent design did not identify when the design occurred either, did it?

"That is correct."

"In fact, intelligent design takes no position on the age of the earth or when biological life began?"

"That's correct."

"It says nothing about what the designer's abilities are?"

"Other than saying that the designer had the ability to make the design that is under consideration, that's correct." As for the designer's motivation, Behe said, "the only statement it [intelligent design] makes about that is that the designer had the motivation to make the structure that is designed."

"How can intelligent design possibly make even that statement, Professor Behe? ... How can it possibly say anything about the intelligent designer's motives without knowing anything about who the intelligent designer is?"

Behe referred him to SETI (the Search for Extra Terrestrial Intelligence), where "the whole project is based on the assumption that they would be able to detect the message without knowing the motives of whoever sent it."

Delving deeper into the possible religious nature of intelligent design, Rothschild read from an article Behe had written: "It is not plausible that the original intelligent agent is a natural entity. The chemistry and physics that we do know weigh heavily against it ... Thus, in my judgment it is implausible that the designer is a natural entity."

Behe conceded that he thought the designer was God, but he referred Rothschild to something he had written in which he asked this question: Is it possible that the designer is a natural entity? "And I won't quote from it, but I come to the conclusion there that, sure, it's possible that it is, but I do not—I myself do not find it plausible. Let me again liken this to the Big Bang ..."

Rothschild pointed out that the year Behe published *Darwin's Black Box* was also the year the Center for Science and Culture was founded. He had been at the Pajaro Dunes conference, and he was a fellow from the start. Had he received money from the Discovery Institute?

Behe said that for about three years the Discovery Institute "gave a sum of approximately $8,000 to $10,000 per year to the university to release me from some teaching obligations so that I could write and think about intelligent design issues."

He had nothing to do with The Wedge strategy document and did not know who had written it, but he had read it, and so was aware that it said, "Design theory promises to reverse the stifling dominance of the materialistic world view and to replace it with a science consonant with Christian and theistic convictions."

"But you continued on as a fellow after seeing this?" Rothschild asked.

"You bet I did. I very much enjoy my association with the Discovery Institute."

. . . .

The next morning there was a conference in chambers, to which, of course, I was not invited. I have since read the transcript of the meeting, however, and its substance was an argument about the admissibility of a new book by the Foundation for Thought and Ethics. It was in draft form, but it seemed to be a redevelopment of *Pandas*, and the plaintiffs' lawyers wanted to be able to use it in court.

A conference call was suggested with Jon Buell, the publisher. Rothschild pushed for doing this at lunch time.

"How much more cross do you have?" asked Judge Jones.

"It will be inversely proportionate to mentions of the Big Bang, I think."

"So you're going to go all day," Jones said.

And indeed he did.

The Big Bang was becoming a universal joke in the courtroom. There was something self-defeating in its constant use. Apart from intelligent design, was this the only scientific theory that had not been accepted because of its possible religious implications—in other words, for nonscientific reasons?

No, there was another, and this was the case of Abbot Gregor Mendel, known as the father of genetics. Mendel was used less often, however, perhaps because it was known that he did not enter the Catholic church for any great religious purpose but "felt compelled to step into a station of life" that would enable him to escape "the bitter struggle for existence."

Mendel published only one article in a single obscure local natural history journal in 1866, which described his experiments on the hybridization of peas. The article and the ideas contained in it were indeed ignored for thirty-five years, until three scientists who were reaching similar conclusions about the characteristics of genetics heard about the article, found it, and read it.

To what extent Mendel's research and conclusions affected theirs is a subject of some speculation, but all three went to great lengths to ensure that Mendel got full credit—more credit, some would say, than he deserved.

Though in some respects they were assumed within his work, "Mendel's Laws" were not actually formulated by him but by his "rediscoverers." One cannot know whether in fact he did more rigorous research and reached firmer conclusions, because when he died, the abbot who replaced him destroyed his gardens and burned all his papers.

As for the Big Bang theory, it was largely "hobbled," to quote Behe, by the astronomer Edwin Hubble, so it might more accurately be described as having been hubbled rather than hobbled. Some scientists supported it from the start, and within thirty years, as more evidence was found, it was generally accepted, albeit provisionally.

Actually, there was a far better example of a scientific theory being "hobbled" for nonscientific reasons than either of these: Darwin's theory of evolution. It was resisted for nonscientific reasons the day it was proposed and was still being resisted a hundred and fifty years later for exactly the same reasons. Big Bang? Thirty years? Ha! A blink of the eye by comparison.

Once again, cross-examination began with the same excessive politeness of the preceding day.

"Good morning, Professor Behe."

"Good morning, Mr. Rothschild."

"How are you?"

"Fine, thanks."

"I'm going to see if we can reach an agreement on something here. You agree that this is a case about biology curriculum?"

"Yes, I do."

"Not about physics, a physics curriculum?"

"It's not about a physics curriculum, but from my understanding, many issues that are being discussed here are particularly relevant to other issues that have come up in other disciplines of science."

"This is a case about what's being taught in biology class, not physics class?"

"As I said, I agree that it is, but one more time, I think many things in the history of science are relevant to this, and they've happened in other disciplines as well."

"You've already testified you're not an expert in physics or astrophysics?"

"That's correct."

"And you might not know this about me, but I'm not either."

"I'm surprised."

"So I'm going to propose an agreement. I won't ask you any questions about the Big Bang, and you won't answer any questions about the Big Bang. Can we agree to that, Professor Behe?"

Muise stood up amidst laughter. "Objection, Your Honor. He's trying to limit the testimony of the witness by some sort of agreement."

After a brief exchange with Jones, the issue was left unresolved, so Rothschild tried one more time, suggesting that surely Behe had "told us everything you know about the Big Bang that's relevant to the issue of intelligent design and biology."

"Well, I'm not sure. I would have to reserve judgment."

"You might have some more?"

"Perhaps."

"Let the record state," said Rothschild with a sigh, "I tried."

Michael Behe smiled. As I've said, I liked him, and to his credit he was good humored and generally imperturbable.

He had, however, published an article, "Tulips and Dandelions," in *Christianity Today,* wherein he wrote, "We want to share the Good News with those who have not yet grasped it and defend the faith against attacks." This statement, when combined with comments that spoke of "materialism" as a weapon used against Christianity, was, as Rothschild pointed out, redolent of The Wedge strategy document.

Behe agreed with a shrug.

In *Darwin's Black Box* he claimed that nowhere in the scientific literature was there a description of "how molecular evolution of any real, complex, biochemical system either did occur or ever might have occurred," and he went on to conclude that "In effect, the theory of Darwinian molecular evolution has not published and so it should perish."

Rothschild pointed out that he, Behe, had failed to publish any of his theories except in his book, and that in the book he had not reported any new data or original research.

Behe agreed, "but I did generate an attempt at an explanation."

Later, trying to explore that attempt, Rothschild returned to the bacterial flagellum. "Did the intelligent designer design every individual flagellum in every bacteria or just the first lucky one?"

"Well, since organisms, biological organisms, can reproduce, of course, then if one has the genes and the proteins and information for a flagellum, then by the normal processes of biological reproduction, more copies of that structure can occur."

"So the answer is, just the first one?"

"That's all that would be needed. That's all we can infer, yes."

Was it possible that there could be multiple designers?

"Yes, I wrote that in *Darwin's Black Box*."

Could there have been competing designers? There could. Was he aware of any irreducibly complex systems that "just came into existence" in the past five years? No. A hundred years? No.

"So scientifically, we can't even state right now that an intelligent designer still exists, correct?"

"That's correct, yes."

"Is that what you want taught to high school students?"

Behe ultimately replied in the affirmative.

If there was a test that could be done to falsify intelligent design, asked Rothschild, or at least the irreducible complexity of the bacterial flagellum, why had Behe not done it? After all, he had himself proposed one that would take a mere two years.

Behe replied at first that actually such a test "wouldn't demonstrate that it wasn't irreducibly complex. It would demonstrate ... that random mutation and natural selection could produce irreducibly complex systems."

"Fair enough," said Rothschild. "It could evolve, and that would falsify your claim that an irreducibly complex system, like a bacterial flagellum, could not evolve through random mutation and natural selection?"

"That's right, yes."

Rothschild asked if *anyone* in the intelligent design movement had performed such a test, and Behe said no. A man named Barry Hall, however, had done similar work on the lac operon without success.

"Professor Behe," Rothschild asked, "you say right here: here is the test, here is the test that science should do, grow the bacterial flagellum in the laboratory. And that hasn't been done, correct?"

"That has not been done. I was advising people who are skeptical of the induction that, if they want to ... come up with persuasive evidence that, in fact, an alternative process to an intelligent one could produce the flagellum, then that's what they should do."

"So all those other scientists should do that, but you're not going to?"

"Well, I think I'm persuaded by the evidence that I cite in my book, that this is a good explanation and that spending a lot of effort in trying to show how random mutation and natural selection could produce complex systems, like Barry Hall tried to do, is likely to result—is not real likely to be fruitful, as his results were not fruitful. So, no, I don't do that in order to spend my time on other things."

To me this was perhaps the most startling admission in the whole trial. Here was Behe advocating that his theory be taught in high schools across America—and yet he could not be bothered to even *try* to test it. Wasn't this the *first* thing you would do? Wasn't this, in fact, exactly what science was?

Behe's attitude reminded me of a remark William Jennings Bryan made in the Scopes trial when questioned by Darrow about the age of the earth.

"You have never in all your life," asked Darrow, "made any attempt to find out about the other peoples of the earth—how old their civilizations are—how long they have existed on earth?"

"No, sir, I have been so well satisfied with the Christian religion that I have spent no time trying to find arguments against it."

Behe, though not explicitly referencing religion, was essentially saying the same thing. By contrast, Charles Darwin had a practice of rigorously writing down everything that might possibly contradict his ideas, because it was so easy, he said, to unconsciously disregard anything that came in conflict with them. He had spent twenty years testing his ideas before presenting them.

A long discussion of blood clotting followed. I don't think any nonscientist in the court understood the details. It seemed that Behe was focused on a narrow aspect of the blood clotting cascade and that he was making the same argument he made with the bacterial flagellum, though perhaps a weaker one because the puffer fish, for example, has a blood clotting system that functions without some parts of the system Behe proposed was irreducibly complex. There were numerous articles in peer-reviewed scientific journals discussing the blood clotting cascade in an evolutionary fashion. Perhaps more complex clotting systems had evolved from simpler ones such as these.

At the end of all this, Rothschild asked, "And before we leave the blood clotting system, can you just remind the court the mechanism by which intelligent design creates the blood clotting system?"

Behe replied: "Well, as I mentioned before, intelligent design does not say a mechanism . . ."

. . . .

Rothschild's last attack on Behe had to do with the immune system. As the arguments were highly scientific and as the essence of them was identical to the bacterial flagellum and blood clotting arguments, I won't bore you with the details. There was, however, one theatrical highlight—an intellectual sight gag, a comic stratagem—which Nick Matzke had suggested to Rothschild back in May. The latter had embraced the idea, and the rumpled science expert from the NCSE had worked tirelessly ever since to assemble the necessary props.

It began with Rothschild quoting from Behe's book: "'We can look high or we can look low in books or in journals, but the result is the same. The scientific literature has no answers to the question of the origin of the immune system.' That's what you wrote, correct?"

"In this context," replied Behe, "that means that the scientific literature has no detailed testable answers to the question of how the immune system could have arisen by random mutation and natural selection."

"Now, you were here when Professor Miller testified?"

"Yes."

"And he discussed a number of articles on the immune system, correct?"

"Yes, he did."

"May I approach, Your Honor?"

"You may."

Rothschild picked up some documents and carried them over to the witness box.

"I'm just going to quickly identify what these articles are . . . 'Transposition of HAT elements, links transposable elements, and VDJ recombination,' that's an article in *Nature* by Zau, et al . . . An article in *Science*, 'Similarities between initiation of VDJ recombination and retroviral integration,' Gent, et al. 'VDJ recombination and RAG mediated transposition in yeast . . .'"

He read the titles of perhaps eight articles.

Behe had told us the day before that he had done a search for the words "random mutation" in the scientific literature dealing with the immune system. He had not, however, as he admitted, done a search for the word "transpositions," which, in this context, according to Rothschild (via Matzke), was more relevant.

Behe disputed this but acknowledged that the articles did seem to be "trying to look at connections between vertebrate immune systems and precursor elements."

Rothschild suggested that the articles in fact rebutted Behe's assertion that the scientific literature had no answers on the origin of the immune system.

"No, they certainly do not. My answer, or my argument is that the literature has no detailed rigorous explanations for how complex biochemical systems could arise by a random mutation and natural selection, and these articles do not address that."

"So these are not good enough?"

Behe looked at Rothschild warily.

"They're wonderful articles. They're very interesting. They simply just don't address the question that I pose."

"And these are not the only articles on the evolution of vertebrate immune system?"

"There are many articles."

"May I approach?"

Rothschild now picked up a much, *much* larger stack of documents—fifty-eight articles on the evolution of the immune system—and hefted them across the court to the witness box and placed them around Behe.

"Is it your position today that these articles aren't good enough; you need to see a step-by-step description?"

"These articles are excellent articles, I assume. However, they do not address the question that I am posing. So it's not that they aren't good enough. It's simply that they are addressed to a different subject."

"You would need to see a step-by-step description of how the vertebrate immune system developed?"

"Not only would I need a step-by-step, mutation-by-mutation analysis, I would also want to see relevant information such as what is the population size of the organism in which these mutations are occurring, what is the selective value for the mutation, are there any detrimental effects of the mutation, and many other such questions."

"And you haven't undertaken to try and figure out those?"

"I am not confident that the immune system arose through Darwinian processes, and so I do not think that such a study would be fruitful."

"It would be a waste of time?"

"It would not be fruitful."

"And in addition to articles there's also books written on the immune system?"

"A lot of books, yes."

"And not just the immune system generally, but actually the evolution of the immune system, right?"

"And there are books on that topic as well, yes."

Rothschild now returned once more to the defense table, picked up a mighty stack of books and staggered back toward the witness stand.

"I'm going to read some titles here. We have *Evolution of Immune Reaction* by Sima and Vetvicka; are you familiar with that?"

"No, I'm not."

"*Origin and Evolution of the Vertebrate Immune System*, by Pasquier. *Evolution and Vertebrate Immunity*, by Kelsoe and Schulze. *The Primordial VRM System and the Evolution of Vertebrate Immunity*, by Stewart . . ."

As he read more and more titles he piled the books up around Behe, hemming him in with contradictory scholarship. In fact, there were so many books that some ended up on Behe's lap for lack of space on the shelflike surround of the witness box.

But still, this was not enough for Behe, who now protested through walls of books: "As somebody who's not working within a Darwinian framework, I do not see any evidence for the occurrence of random mutation and natural selection."

"Let me give you some space there," said Rothschild, removing a portion of Behe's prison wall.

"Thank you."

Rothschild dutifully lugged all his material back to the defense table but returned with another book in manuscript form.

It was the new version—or perhaps "new model"—of *Pandas,* which, it seemed, was now to be called *Design of Life.*

The main force behind this, it appeared, was William Dembski, who had, apparently without asking Behe's permission, already listed him as a contributor.

Within the book was a new expression that Matzke had discovered a few months before and found amusing at the time. He had brought it to the attention of the lawyers, but in the plethora of other matters, all had forgotten it until this very morning. As Nick and Vic were roommates, they often drove to court together. Now, as they approached the court, the latter had a dim recollection of the phrase, which Matzke rapidly confirmed.

When they reminded Rothschild of its existence, he quickly took hold of it and used it as the punch line of his cross-examination.

Just as creationism had been replaced by creation science, which in turn had given way to intelligent design, the new book intimated that the latter might soon seem just too passe for next year's high school, and so required replacement with a snappier sobriquet. In the chapter entitled "The Fossil Record," *Pandas* used the following phrase:

"Intelligent design means that various forms of life began abruptly through an intelligent agency, with their distinctive features already intact . . ."

In the manuscript for the newer, fresher, brighter book, the sentence read:

"Sudden emergence holds that various forms of life began with their distinctive features already intact . . ."

"Hopefully," said Rothschild, "we won't be back in a couple of months for the Sudden Emergence trial . . ."

"Not on my docket," said Jones, "let me tell you."

• • • •

Having watched Michael Behe for many hours, I was, at the end of it, unsure how much he himself believed what he was saying. His philosophy rested on "the argument from ignorance" or "the God of the Gaps." Whatever in the universe appears beyond our comprehension, whatever gaps remain in our knowledge, must be ascribed to God.

This is a form of arrogance. It was as if Behe was saying, "If I cannot understand this puzzling feature of the universe, no one ever will."

If you look at God in this way, his habitat has clearly shrunk. To list the gaps from which he has been ejected is redundant. Darwin was the author of the most comprehensive eviction notice and clearly the chief villain of the piece.

In Behe's chapter in *Of Pandas and People,* he writes: "Like a car engine, biological systems can only work after they have been assembled by someone who knows what the final result will be."

But the car analogy seems, at least in the context of irreducible complexity, to offer a better argument for evolution than intelligent design. You could not "see" the car evolving from a single object—a log or a boulder, which at some distant time someone had rolled for the purpose of moving something from one place to another, which had in turn evolved into a wheel—but you could certainly infer that this was how it began, and you could certainly see the car's evolution from a horse-drawn cart to its present form.

To argue that because a car cannot function without its steering wheel it must therefore have had a steering wheel in all previous incarnations is clearly nonsensical. While it is true that an intelligent designer was involved in the creation and evolution of the car, most people believe that its designer—the human race—evolved its intelligence through natural processes.

The car analogy is weak on another level. The car sprang from the rock or the log in the blink of an eye, a few thousand years, while evolution has had, according to best estimates, three and a half billion years to bring life to its present form.

This is the central barrier to understanding evolution. We understand time through the experience of our own short lives. To truly imagine three and a half billion years is virtually impossible. Imagine yourself living to seventy—I mean really *imagine* seventy years: being born, a decade and a half of education, many more decades of employment, wars, elections, scientific discoveries, parents lost, middle age, old age—innumerable memories marked off by seventy birthdays and seventy summers and winters.

Now try to imagine fifty million of those lifetimes—*fifty million* of them! Because that is how long life has been developing on earth.

But how can you begin to conceive of such an expanse of time?

Try this. If, at a modest clip—which I'd recommend, given what I'm proposing—it takes you a minute to count out loud to a hundred, it will take you almost a week of nonstop counting to reach a million. That is, counting without a single break and no sleep.

If you could keep counting for twenty-four hours a day for three hundred and fifty days, you'd reach fifty million. But these are not just meaningless numbers—each one of them represents a lifetime. But almost a year without sleep is inconceivable, so let's try and make this "doable," as Behe would say.

Put in eight hours of counting a day, seven days a week. Take a two-week vacation each year.

Under these still-harsh working conditions (no weekends off), it will now take you *three years* to count out these fifty million lifetimes.

(You will reach, incidentally, the birth of Christ within the first half minute, and the oldest age of the earth, according to believers in a literal Genesis, within the first two minutes.)

But to really comprehend this expanse of time, you would still have to be capable of imagining—as each of those numbers came tripping off your tongue, hour after hour, week after week, month after month, year after year, for *three years*—that each of those numbers signified *a lifetime*.

Even if you chose to do this, and even if you were capable of the extraordinary effort of will and imagination needed to conceive of what you were actually doing, I suspect that at the end of it you would still be only a little closer to comprehending the vast amount of time involved.

In all probability, you would give up long before you finished, overwhelmed by depression at your own insignificance. It is offensive to one's sense of self to imagine this huge expanse of time that came before you and within which you had no relevance.

No, it is more than offensive; it is terrifying. How much easier—and how much more comforting—to just put in those first two minutes and imagine, in one way or another, a designer who placed you at the center of it all.

Moral Decay

IF I BELIEVED that someone was watching over me and that when life ended, death would be a joyous reunion with everyone I loved, I would be ecstatic. Why, then, are so many true believers so often consumed with rage and bitterness?

While I was down in Harrisburg, I met a man who claimed to believe absolutely that God existed, that he created the world in six days, and that there was a heaven to which he would go. His whole life was based on these beliefs, and yet everything about him was angry and mean-spirited.

Reverend Jim Groves was a fixture at the trial, sitting in the back watching the proceedings through narrowed eyes. One night he put on a show in Dover. Advertised as *Evolution Is Stupid*, it was held in the Dover firehouse and began with a showing of a DVD by that name made by a man named Kent Hovind, a former biology teacher.

Hovind, a man who knew enough science to be able to poke fun at it in the right environment—one populated by believers of the lowest type—would throw up an aspect of evolutionary theory, couched in words complex enough to be baffling to his audience, and then took shots at it with the phrase "That's stupid!" or "I'm sorry, boys and girls, but that's not common sense; that's just stupid!"

When this endless clay pigeon shoot was done and the DVD was turned off, a hero of mine, Burt Humburg, a medical intern at Penn State, calmly raised up a tabletop document stand and began to defend evolution.

Within moments, a woman whose fangs had manifestly dodged the modern science of dentistry stood up to yell out the battle cry of the fundamentalist: "You've been brainwashed in college! You've been brainwashed in college!" There were murmurs of agreement, and Burt, though he struggled on manfully, was soon silenced. I caught up with him later and found, to my increased admiration, that he had been raised in a charismatic division of the church where the believers spoke in tongues.

A few days later I interviewed Groves for my documentary. A wiry little homo-hater in his late fifties, dressed in tight pants and cowboy boots, he had an insinuating manner that belied his courage: every Halloween he insisted on joining the parade in York, putting on one of those gruesome antiabortion shows so beloved of the breed, smashing blood-filled dolls and displaying graphic photographs of aborted fetuses, and so scaring the children that one year he was banned. He appealed, however, on the basis of free speech, won, and was now a fixture, albeit at the rear.

By this time, it was public knowledge that I was an offspring of Darwin, and in the course of the interview I became viscerally aware of just how comprehensively hated the poor old boy is. People such as Groves believe that Darwin marks a point in history which, if it wasn't in use already, would usher in a dark age described as A.D., or After Darwin. Hitler, Stalin, Pol Pot, drugs, sex, prostitution, abortion, homosexuality, violence, suicide—all the failings of modern society were ascribed directly to Darwin.

I asked him why he thought Darwin and all those who followed in his footsteps believed in evolution. Why would they make this up?

Groves replied that it was because they wanted to escape God. If there was a God, he held you accountable; if there wasn't, you could do what you wanted. "You see where our culture's going, with sodomy, gay marriage; this is a result of taking off the restraints which God has put on the moral code ... When you believe evolution, it affects everything you believe. I've got a seminar I do on evolution ... I've got ten different areas that evolution affects. It affects laws, it affects politics, it affects sociology, it affects psychology."

He referenced pictures in Miller and Levine's *Biology* showing "where we in the womb supposedly go through the embryonic stage of a fish and a mammal and a reptile ... and hey, if you're gonna have an abortion, it may not be a person yet. It might be a fish. It might be some small mammal. It might be a reptile. So now you can logically take that and say, 'Well, you know, it's just not a child.'"

Most of the conflicts between evolution and the bible concern the first few chapters of the latter, and it was on these that Darrow had questioned Bryan at the very end of the Scopes trial. I decided to ask Groves some of the same questions to see whether anything had changed.

Bryan, when pushed, was an old-earth creationist—the six days of creation could have been "ages" rather than literal days, and the earth might be older than the eight to ten thousand years suggested by the bible.

Groves believed that the days were literal and the earth could not be more than ten thousand years old. He based this on the bible itself, the mathematical calculations of William Dembski, and tests that showed carbon dating was nonsense.

While down in Dover, I discovered that this approach was fairly typical, an odd jumble of biblical faith mixed in with scraps of information from the scientific fringe, much of which was clearly incomprehensible to the believer. In fact, I came to see that it was often the incomprehensibility of these "scientific theories" that made them plausible. They sounded "scientific," therefore they were.

Was Jonah really swallowed by a whale?

Both Groves and Bryan believed he was.

Darrow asked Bryan whether he believed that "Joshua commanded the sun to stand still for the purpose of lengthening the day."

Bryan had answered: "I do."

Darrow pushed. "Do you believe that at that time the sun went around the earth?"

"No, I believe that the earth goes around the sun."

"Have you an opinion as to whether whoever wrote the book—I believe it is Joshua, the Book of Joshua—thought the sun went around the earth or not?"

"I believe he was inspired."

"Can you answer my question?"

"I believe that the Bible is inspired, an inspired author. Whether one who wrote as he was directed to write understood the things he was writing about, I don't know . . . I believe it was inspired by the Almighty, and He may have used language that could be understood at that time instead of using language that could not be understood until Darrow was born."

Groves was less evasive.

"Sure, I believe it happened."

"But, if the earth is going around the sun, how can the sun have stopped in the sky?"

"Well, I understand what you're saying there. And the earth rotates, of course, okay? And that's basically what happened. The earth stood still, you know? I mean, it came to a stop to lengthen the day, you know?"

"So, he didn't stop the sun in the sky, he stopped the earth rotating?"

"Yeah."

"But, then what would happen to gravity?"

Groves responded that as God was the author of the laws of nature, He could overrule them.

Were all living things not contained in the ark destroyed?

Bryan had replied: "I think the fish may have lived."

Groves forgot about the fish but thought some plants might have survived.

Was the story of Adam and Eve to be taken literally?

Both Bryan and Groves thought it was the simple truth. That is how man came into being.

"And this is a question," I asked with some trepidation, "that I know is going to get on your nerves. Where did Cain get his wife?"

But my trepidation was unwarranted. Groves was happy to explain. Cain married his sister or niece. "And that process was outbreeding, okay? God put a limit on marrying relatives later because that now becomes inbreeding, okay?"

I left it at that, not having the nerve to go as far as Darrow had with Bryan.

"Do you believe," Darrow had asked, "the story of the temptation of Eve by the serpent?"

"I do," replied Bryan.

"And you believe that is the reason that God made the serpent to go on his belly after he tempted Eve?"

"I believe the Bible as it is, and I do not permit you to put your language in the place of the language of the Almighty. You read that Bible and ask me questions, and I will answer them. I will not answer your questions in your language."

"I will read it to you from the Bible: 'And the Lord God said unto the serpent, Because thou hast done this, thou art cursed above all cattle, and above every beast of the field; upon thy belly shalt thou go, and dust shalt thou eat all the days of thy life.' Do you think that is why the serpent is compelled to crawl upon its belly?"

"I believe that."

"Have you any idea how the snake went about before that time?"

"No, sir."

"Do you know whether he walked on his tail or not?"

The audience laughed uproariously.

Darrow had more questions, but as he was in the middle of asking the next one, Bryan launched off into one of his many martyred protestations.

"Your honor, I think I can shorten this testimony. The only purpose Mr. Darrow has is to slur at the Bible, but I will answer his question. I will answer it all at once, and I have no objection in the world. I want the world to know that this man, who does not believe in God, is trying to use a court in Tennessee—"

"I object," said Darrow.

"—to slur at it, and while it may require time, I am willing to take it."

"I object to your statement," said Darrow, losing his temper. "I am examining you on your fool ideas that no intelligent Christian on earth believes!"

Suddenly the two men were on their feet, glaring at each other.

Everyone stared at them in shock. Was Bryan going to hit Darrow or Darrow Bryan? The judge banged his gavel, and the day was over.

Darrow emerged a hero. Bryan had been made to look like an ignorant fool.

Five days later, while still in Dayton, Bryan died while taking a nap after lunch. Some thought he had died of a broken heart after his mistreatment by Darrow. When this hypothesis was put to Darrow, he said, referring to Bryan's well-known gluttony, "Nonsense, he died of a busted belly."

H. L. Mencken had a different theory. "God aimed at Darrow, missed, and hit Bryan instead."

• • • •

Fundamentalists dream of a return to a golden age of America before the stifling dominance of materialism ruined everything, a time when Christianity ruled supreme. As Reverend Groves said, "We no longer have religious freedom in America; we have a religious *free for all* in America. America was not that way, was not that immoral when it stayed to its Christian roots."

But where is this golden age? Would pre-Darwinian America look so golden if we had documentary footage of the slaughter of American Indians, or could watch the importation of slaves on the nightly news? What kind of enlightened morality permitted the hanging of "witches," not to mention children as young as eleven, both of which occurred in America long before *The Origin of Species* was published?

It is true that before Darwin, few Christians had any particular reason to doubt the literal truth of the bible, and that after him, Genesis, at least from a literal standpoint, looked ridiculous. Men like John Haught refined their view of scripture and found a way to extract a deeper meaning from it that was not in conflict with scientific truth. Fundamentalists simply dug in their heels and obstinately denied the evidence.

When I arrived in his small, depressing basement church, Groves had already set up his own little video camera to film me filming him. He told me it was just to "keep a record of the event," and I did not object.

At the end of my interview, he asked what my religious beliefs were, and I told him. I thought faith was understandable but unhealthy, and that consequently I was not an atheist but an agnostic, because having faith, even in nothing, was too much faith for my taste. I then asked why, if Christian faith was so marvelous and America had so much of it, poverty, sickness, crime, and imprisonment were so rampant and so ignored.

I did not get a satisfactory answer, but a couple of days later, two reporters told me they had visited the church in search of local color and found me booming from a TV on the altar, declaring my agnosticism to many gasps of horror. Apparently, the consensus was that I'd end up in hell, probably to find grandpa sitting at the devil's side. When I upbraided Groves about this—he had not told me I was to be used in this way—he shrugged my objections aside and told me it had been "educational." He and his flock had eventually concluded I had a different understanding of

Christianity. Coming from Europe, mine was "more socialistic," while his was more concerned with "individual salvation."

This view of Christianity, or so it appears to me, is exactly the problem, encouraging as it does an entirely selfish view of life—and indeed the afterlife—that can currently be detected in every mean-spirited policy that emanates from the White House.

As I have read the bible more assiduously than *The Origin of Species*, accounts of the life of Jesus, though containing some fairly vicious comments, do not seem to support Groves's view, and you can't help but wonder whether a lot of Christians never actually get past the Old Testament.

Sudden Disappearance, Sudden Emergence

THE PLAINTIFFS PRESENTED six expert witnesses. The defense eventually mustered three. Apart from Michael Behe, none of the leading lights of intelligent design appeared. Not Phillip Johnson, not William Dembski, not Steve Meyer.

Both Dembski and Meyer had been on early expert witness lists provided by the Thomas More Law Center, but they dropped out when Thompson refused their demands to have their own lawyers present during depositions. According to the Discovery Institute's website, Thompson later relented and said Meyer could bring a lawyer. By that time, however, Meyer had "lost confidence in their [the Thomas More Law Center's] legal judgment."

The rift between the Discovery Institute and the Thomas More Law Center would culminate in a public argument at a forum in Washington sponsored by the American Enterprise Institute. Both Mark Ryland, director of the Discovery Institute's Washington office, and Richard Thompson—taking a break from the trial—were present.

Ryland explained that Dembski and Meyer had wanted lawyers because they had "different legal interests." Thompson essentially accused the Institute of cowardice, saying, "as soon as there's a controversy, they back off with a compromise."

Ken Miller, the Catholic biologist from Brown University and the plaintiffs' first expert witness, was also at the forum and could not contain his glee as the two sides lashed out at each other. "I'm really enjoying this," he said during a break in the argument. "That is the most fascinating discussion I've heard all day!"

When Behe finished on Wednesday, the defense said their next expert would be a man named Dick Carpenter. Dick was an assistant professor of "leadership, research, and foundations" at the University of Colorado at Colorado Springs. He was also active in "Love Won Out," a Focus on the Family program designed to "cure" gays.

The defense did not want to put him on until Friday and seemed a little confused as to his whereabouts. Patrick Gillen said, "We do have an expert coming in," but when the judge asked, "The expert is not going to get here until Friday?" Muise replied, "No, he's been here since last night." "So we want to get him in and get him out of town," said Gillen. But instead of putting him on right away, they'd put Richard Nilsen, the school superintendent, on, then put on Carpenter, then return to Nilsen.

"You don't want to start with the expert?" asked Jones, clearly baffled.

"Right," said Gillen. "We'll start him on Friday and get him done, and then get him out of town because he needs to get—"

"You're going to get him done on Friday, though?"

"He's a short expert from our perspective," said Muise.

He was. Extremely short. He never appeared in court at all, for reasons that never became clear.

Instead, we were treated for the next day and a half to the interminable and mind-numbingly dull testimony of the school superintendent. Only the following Monday did we get "an expert," a man named Stephen Fuller, who flew in from England.

One couldn't help getting the feeling that the intelligent design rats were leaving the sinking ship—the smarter ones, like Dembski and Meyer, leaving it early, perhaps to concentrate on evolving into Sudden Emergence rats.

Richard Nilsen, a narrow-faced man in glasses, was, according to Carol Brown, the son of missionaries plying their trade in Africa. He gave his testimony in a monotone so utterly devoid of color or expression that after the first day I suggested to some fellow reporters a "Nilsen Poetry Reading Contest" and composed some guidelines.

"The prize will go to whoever reads or recites the most passionate poem in the most consistent monotone. Particular note will be taken of your ability to extract any trace of humanity from your poem. Participants should avoid wearing colorful clothes or smiling."

The prize was won at the Indian restaurant next to the Comfort Inn by a famous author and reporter from Texas. I think he recited a poem by Robert Frost, but his recitation was so spectacularly dull that I have no memory of it whatsoever.

I have little memory of Nilsen, either. Suffice it to say that he was a manikin assembled in homage to bureaucracy and varnished with a sheen of resentment. Too cruel, you say? Perhaps, but this was a man who could, had he wanted to, have slammed the door on Bonsell and Buckingham and avoided this whole expensive mess. Although the board had the power to fire him, he was, nonetheless, "the superintendent," *Doctor* Nilsen, an educated man, educated in education in fact, a man in possession of a doctorate in education administration.

True, he got this from Temple University, founded in the late nineteenth century by Russell Conwell, an evangelist preacher popular on the lecture circuit for his "Acres of Diamonds" speech. I assumed the title was lovingly ripped off from William Jennings Bryan's "Cross of Gold" speech, until I read it.

"Cross of Gold" was a rousingly beautiful and poetic defense of the working man. "Acres of Diamonds" was a robust defense of Rockefeller and Carnegie and an attack on the poor and on unions that "scale down a man that can earn five dollars a day to two and a half a day, in order to level up to him an imbecile that cannot earn fifty cents a day."

The speech makes fascinating reading because it appears to be one of the earliest iterations of "prosperity theology," so popular now with pentacostalists and televangelists of all stripes.

Superintendent Nilsen endeared himself to me conclusively when he was asked by Gillen why he'd never done anything about press coverage of school business that was, according to him, inaccurate. With a curl of his lip, he said he had learned long ago to abide by the adage "that you never take on individuals that buy ink by the barrel."

Judge Jones glanced at me and raised his eyebrows as if to say, "I think he's talking about you, buddy."

It was such a funny and apposite gesture that I had to look down in order not to laugh. When I learned that the quote (more accurately, "never pick a fight with people who buy ink by the barrel") was attributed to Mark Twain, I was even more amused.

It was Twain, after all, who wrote, "Faith is believing something you know ain't true," that the bible is "perhaps the most damnatory biography that exists in print anywhere. It makes Nero an angel of light and leading by contrast," and that if God existed he was "a malign thug."

Not exactly the source you'd expect the son of missionaries to quote from.

Such a consummate bureaucrat was Nilsen that after many hours of testimony, his motives remained a mystery. Was he afraid of getting fired? Almost certainly (and this was a bullet he would not ultimately dodge). Did he actually approve of intelligent design? Probably. One thing you could say: if lack of curiosity was the hallmark of the intelligent design supporter, he was on that side. In spite of the potential financial consequences for the district, he seemed to have done almost no research into either intelligent design or evolution.

A few days later, Michael Baksa, the assistant superintendent, took the stand. He was a likeable man and somewhat more candid but, like his boss, revealed little. I sensed, not just from watching him at the trial, but from talking to teachers and board members, that he was a decent man embarrassed by the whole thing. He had, most people guessed, simply been unable to stand up to Buckingham, Bonsell, and Nilsen.

A couple of weeks after the judge made his ruling, Baksa crashed his car. He survived unhurt but was found to be drunk.

. . . .

Steve Fuller, the suddenly emergent expert who was rushed into the schedule to replace the oddly gone-missing Dick Carpenter, was a supercharged man full of superlatives.

He was currently a professor of sociology at the University of Warwick in England, "one of the top five research universities in Britain." He graduated from Columbia University, "summa cum laude" in 1979 and then went on to get an MPhil from Cambridge and a PhD in history and philosophy of science from the University of Pittsburgh, which had "probably the best department, certainly in the United States, and probably in the world." He had "200 published articles or book chapters, the vast majority of which have been peer reviewed," and he was "the first National Science Foundation post-doctoral fellow in history and philosophy of science in 1989."

Fuller was fuller of himself than any other expert witness, and not without reason. He was a pleasant, highly intelligent, goofy-looking fellow with large rubbery lips that moved at an astonishing pace as if he might not have time to tell the court of all the wonderful things he had achieved. He spoke so fast that the judge had to tell him to slow down because the court reporter couldn't keep up with him, and when the morning break was taken, Jones nodded at him and said, "Water or decaf only."

What Patrick Gillen was clearly looking for from Fuller was full approval of intelligent design. What he got, ultimately, was a fascinating discourse on, among other things, the religious motivations for scientific research throughout history; a highly nuanced political analysis of the current state of science and how it kept new theories at bay; and the suggestion, really emanating out of the first of these, that intelligent design was an interesting way of imagining how life might have developed, and should not be banished merely because it had not yet produced anything in the least bit useful.

Americans had more respect for science during the Cold War, he told us, because the war was defined in part as a science race between the United States and the Soviet Union. Since America had won the war and become the dominant force in the world, science no longer had the same national security value, and respect for it had diminished along with government funding.

During the Cold War, the government not only provided money for science but also acted as an arbiter of what was worthwhile. Now that imprimatur was provided more by peer review. This was how you achieved academic credibility and thus funding. But as each branch of science became more specialized, the number of peers qualified to approve your work diminished. This led, in Fuller's view, to a kind of "self-perpetuating elite," which could act as a "bottleneck" to shut out new ideas.

He believed that I.D. was science, that it was not religious, and that it was as testable a theory as evolution. All it required was a scientific revolution, because at the moment there was a "dominant paradigm" which stood in its way.

Evolutionary theory was, in a sense, "a kind of universal rhetoric of biology," but if you actually searched for "evolution" or "natural selection" in the scientific literature of biology, you found that those words occurred in no more than 10 percent of the articles. (This suggests to me that evolution is so universally accepted you don't need to keep referring to it.)

Darwin, according to him, was "more like a Copernicus figure . . . he kind of makes the big intellectual change, but he doesn't really provide a basis for research." By the first third of the twentieth century, genetics was the ascendant biological science and seemed at times to contradict various Darwinian theories. Darwinism was "seen as a kind of, you know, old-fashioned natural history—guys who like to look at animals and plants and give 'Just So' stories about how they managed to survive," but it had no clear sense of the mechanism by which this happened.

Not until a man named Theodosius Dobzhansky came along in the 1930s did the two sides came together. Dobzhansky was both a natural historian who did field work and someone experimenting in the genetics lab at Columbia University. It was he who introduced "a common rhetoric," "a language of mechanism" that both sides could agree on.

In other words, the "neo-Darwinian synthesis" was the result of "a big tent theory."

"Well," said Patrick, "given what you said about the situation with respect to the neo-Darwinian synthesis, would you expect the situation to be different for intelligent design theory?"

"No, not at all," replied Fuller, which sounded great, except that he immediately went on to say, "And, in fact, I think, you know, the main problem intelligent design theory suffers from at the moment is a paucity of developers . . . so what you don't have is really a lot of room for theory development, for developing the terms of the argument, and for developing research programs in the area."

Did he see intelligent design as being necessarily linked to natural theology and people such as William Paley?

Fuller replied that the problem with intelligent design was that it didn't know its own history. "It hasn't yet recovered all of the intellectual roots that, in a sense, could provide sustenance for it." If he were offering advice to I.D. people he would go to Sir Isaac Newton. "He is the 400-pound gorilla of intelligent design theory, because this is the man who quite clearly thought he got into God's mind and figured out the basic principles by which all of physical reality was governed."

Using William Paley as "the poster boy for intelligent design" was actually a mistake, because he was defending design as opposed to Newton, who had, in a sense, *done* design. Although elsewhere he was not flattering of Robert Pennock, he felt that one of the most promising areas in which you might "do" design, or think like God, was in the kinds of experiments Pennock was doing with computer-generated organisms.

Gillen asked him if new research was a necessary ingredient of scientific progress, to which Fuller replied that eventually it was, but that all theories needed to reach a certain "critical mass of theory and interpretation of data." It was well to remember that "all new theories are born refuted" because there was always some dominant paradigm that they had to overturn.

Fuller was master of the back-handed compliment and the deft spin of faint praise. The following exchange was typical.

Patrick served him a question: Did he think the criticism that intelligent design had failed to produce experiments disqualified it from science?

"No," Fuller said, smiling reassuringly.

"Why is that?"

"Well, I mean, it's too young basically at this point. And it hasn't really done all of the theoretical elaboration or the recovery of the appropriate history to set itself in a proper tradition that then would kind of field the imagination to come up with the right kinds of experiments."

This was essentially to say that intelligent design was not just half baked; it wasn't even aware of what ingredients it needed to mix together on the counter top. And yet, to continue the food metaphor, its advocates were already trying to force what few uncooked and often old ingredients it did have down the throats of innocents. To argue that "thinking like God" might be a wonderful spur to the imagination was terrific. To cause indigestion in children was not. Worse than irresponsible, it was lazy.

When asked if he thought intelligent design was testable, Fuller spoke of evolution and the extent to which it could not be tested, concluding eventually that in a sense, neither evolution nor intelligent design could be directly tested. The difference was that "evolutionary theory is much more developed, much more elaborate, and in that way, much more suggestive of forms of research to do, which then, in turn, can be tested. So it's got that advantage." Intelligent design, by contrast, was caught in the bottleneck of peer review and could not get funding for its research. The deck was "stacked against radical innovative views."

Did this, in part, explain why he was supporting "Dover's small step in this case?" asked Gillen.

It was, "because if you think about this strategically, how do you expect any kind of minority view with any promise to get a toe-hold in science? Okay. And you basically need new recruits. This has been the secret of any kind of scientific revolution or any kind of science that has been able to maintain itself."

The idea that the kids I met outside Dover High were going to be recruited into the vanguard of a scientific revolution was hilarious. An image flashed into my mind of Johnson, Behe, Dembski, and Meyer lurking in the school parking lot with a bunch of ecstasy pills stamped with the letters I.D.

"Pssst, take this and join the revolution!"

But in a way it was not so funny, because this is exactly what *was* happening. To continue the food metaphor, the school was trying to feed their kids the best food they could find under current understandings of nutrition, while those in the intelligent design crowd were trying to push an untested drug with no apparent benefits.

. . . .

Vic Walczak's opening question in a way addressed this thought. He asked Fuller the meaning of "heuristic," and Fuller replied, "something that helps you imagine situations so that you can come up with hypotheses in science."

Walczak then read him one of the paragraphs of the famous Dover High statement: "Because Darwin's theory is a theory, it continues to be tested as new evidence is discovered. A theory is not a fact." As a theory never became a fact in science, wasn't this misleading?

Fuller waffled. Walczak took him back to his deposition, wherein he had spoken more simply to the issue by saying that this particular phrase in the statement "did seem to want to denigrate" the word "theory" and give more value to the word "fact" when the former was actually more significant than the latter.

Walczak referred Fuller to the statement's concluding sentence: "As a standards-driven district, class instruction focuses upon preparing students to achieve proficiency on standards-based assessments."

In his deposition, Fuller had agreed this was "a downbeat ending because what we should be doing is trying to encourage students that science is fascinating and interesting," rather than implying, as did this lugubrious call to drudgery, that it was something one "had to do" in order to pass a test. Or, to continue in the vein of this food-metaphor–laden chapter: "Eat your broccoli!"

"But," said Fuller in court, "if this is going to be the only way they can actually end up allowing intelligent design as a possibility, then one lives with it."

Walczak then established that Fuller was not familiar with Miller and Levine's *Biology* book or with *Pandas* and had no expertise in biology, paleontology, education (science or otherwise), Behe's irreducible complexity, or Dembski's specified information theory.

At a certain point, Fuller said, agreeing to all this, "I try to be modest."

His approach to this case was, Walczak suggested, purely philosophical. As a philosopher he just "wanted to keep a more open mind on the rules of science."

"I don't know if I'd exactly put it that way, but let's say I certainly warm to that suggestion."

In short order he conceded that he believed evolution provided a better explanation of biological life than did intelligent design—"at the moment"; that compared to evolution's explanation of biological life, intelligent design had not been able to develop itself to be in a position where it could "make any kind of definitive judgment"; and that his commitment to the doctrine of intelligent design was conditional: "I want to see where intelligent design goes, frankly ... you have to give it some time to develop, and then you can make further judgments whether it was a good bet or not."

As for intelligent design being a supernatural idea, "I mean supernaturalistic just means not explainable in the naturalistic terms. Right? It means involving some kind of intelligence or mind that's not reducible to ordinary natural categories. Okay? ... I'm not saying, you know, they're committed to ghosts or something. See, I'm not sure what exactly—but that's how I—I understand supernaturalistic in this fairly broad sense."

In his deposition he had said, "I'm not doubting that methodological naturalism has worked for science and that it's largely responsible for lots of science that we've got, maybe even most of what we've got." Nonetheless, he now believed that the ground rules of science had to be changed in order to admit intelligent design. Methodological naturalism was not "a ground rule of science"; testability was.

"And how do you test the supernatural?" asked Walczak.

"Well, that's an age-old question, but there have been paranormal experiments ..."

At times he seemed to be confusing "theory" with "hypothesis." He said, "No theory is born well substantiated." In fact, using the scientific definition of "theory," a theory is not a theory *until* it is well substantiated.

He kept returning to the fact that intelligent design had not been able to make progress because it was being resisted for ideological reasons by a structure that it was dependent upon for financing.

Walczak pointed out that dozens of controversial scientific theories had eventually achieved acceptance. Was he really saying that intelligent design was so uniquely victimized "that the only chance it has is for a federal judge to order that it be taught in the schools?"

Under redirect, Gillen asked Fuller whether he believed intelligent design was capable of advancing to a stage where it could engage in experimental work.

"It still needs to be developed a bit more," said Fuller, " but in principle it could. But it really does need more adherence and more time to sort of develop the implications of its views."

• • • •

One of the "implications of its views" would have to be that it might actually discover an intelligent designer, or God by any other name.

If lack of money was all that stood in the way of this, I wondered why Howard Ahmanson and Pizza Man had not put everything they possessed into the search. It could, after all, lead to the greatest scientific discovery of all time. To finally find God? What price could be too high? Imagine getting this call from Behe and Dembski:

Behe: "Okay, we found Him."

Pizza Man: "Is he Catholic?"

Behe: "No, sir, I'm afraid he's not."

Pizza Man: "Five billion dollars and he's not Catholic!?"

Ahmanson: "Is he Protestant?"

Dembski: "I'm afraid not. He's a gay Jewish atheist, sir."

Ahmanson: "Gay?!"

Behe: "He said we should have figured it out. The virgin birth? Send an angel to do a man's job? Jesus? Confirmed bachelor, hung out with twelve guys . . ."

Pizza Man: "But why an atheist?"

Dembski: "Said he lost faith in himself after we did the Inquisition, sir. In retrospect, I suppose these were all logical inferences we could have—"

Et cetera.

But seriously—if I may use that disingenuous phrase to slip out of comedy for a moment—if they really *do* believe in the idea, why *don't* they *really* fund it?

I have two hypotheses—or, to be fair, two hunches, because I can think of no way of proving either.

Hunch One is that if you observed these men, they often had the bitter, defiant mannerisms of high school losers. It was easy to imagine them getting knocked around by witless hedonistic jocks in the hallways of whatever schools they attended. This swaggering force of athletic superiority was the unconquerable "dominant paradigm" of their childhoods, and now, for all their wealth and power, when they saw another "dominant paradigm"—other than wealth and power—they instinctively wanted to knock its teeth out. In this, they reminded me of Cheney, Rove, and Rumsfeld, all of whom have that quality of runtish boys who demanded to be beaten up at school and who, now in power, relish every chance to beat us up or, better yet, get us to beat each other up.

The hatred men like Ahmanson and Pizza Man have for evolution has, I am suggesting, as much to do with the smugness of its unconquerable practitioners as with its ideas. Evolutionists are the jocks of academia, and anything that wipes the smiles off their arrogant faces is worth some effort.

Some effort. These men, for all their faith, are not fools. They enjoy—and this is where I find myself in spiritual accord with them—rattling the bars of any cage within which a smug pedagogue lives. What they aren't willing to do is bet a lot of money that these pedagogues are actually mistaken.

Hunch Two is this: These *are* devoted Christians and have read their bibles from cover to cover. What if intelligent design did prise God out of his shell, and what if the old boy was so enraged at being awakened from his long miracle-free hiatus that he lashed back at them by reiterating some of the more controversial aspects of the good book, such as, per Jesus, "It is easier for a camel to go through the eye of a needle than for a rich man to enter into the kingdom of God."

In other words, they might love God but he might not love them back—and who would spend money to find that out?

Buckingham

OUTSIDERS SUCH AS myself were in a froth of anticipation for the testimony of the pugnacious, once–OxyContin-addicted crusader, Bill Buckingham. What kind of brute would this be? But when he arrived, walking with a cane, he seemed tired and subdued. He had been through two stints of rehab to kick his addiction to OxyContin, and I wondered whether another drug had been prescribed to stop him from making outrageous statements in the courtroom.

He testified in a low, mildly surly voice and unashamedly stated that he believed in a literal reading of the book of Genesis. It was clear he knew almost nothing about evolution except that it "happened by chance." As for intelligent design, he seemed to know no more about it than he had during his deposition. When Steve Harvey had asked him to offer a definition, he gave the following response: "A lot of scientists, don't ask me the names, I can't tell you where it came from, a lot of scientists believe that back through time something, molecules, amoeba, whatever, evolved into the complexities of life we have now."

Steve Harvey, astonished that he would give this definition—a definition of evolution, not intelligent design—asked, "That's the theory of intelligent design?"

"You asked me my understanding of it," replied Buckingham. "I'm not a scientist. I can't go into details and debate it with you."

In court, more than nine months later, he said, "I just know that it's another scientific theory that we thought would be good to have presented to the students."

He admitted having made the remark about Christ on the cross and people taking a stand for him, but claimed it was at a meeting a year or two earlier when the board was debating the Pledge of Allegiance.

The most important part of Buckingham's testimony, and the source of one of the most dramatic moments in court, was his (and later Bonsell's) contention that they had not used the word "creationism" in the afore-reported school board meetings but had been fixed on the scientific theory of intelligent design from the start. Their intent had never been religious. Just as reporters (and many other witnesses) had lied about his Christ on the cross remark, so too had they lied about this.

But, if this was the case, asked Steely-eyed Steve, why hadn't he and Bonsell objected to the newspaper reports at the time? Buckingham explained that as he didn't believe "a darned thing" they printed. He opened the newspapers only to "read the obituaries and see how my fighting Phils did." He had no idea he was being quoted as using the word "creationism."

A few minutes later, Harvey asked, "Now, it's your testimony that at neither meeting no one on the board ever mentioned creationism, isn't that right?"

"That's true."

"You're very clear on that point, correct?"

"Absolutely, because it's just something we didn't do."

Harvey asked him if he'd mind looking at exhibit P–145.

The Wizard of Oz tapped a few buttons, and the screen came to life. Here was Buckingham being interviewed by a local TV news reporter outside the school.

"The book that was presented to me," he calmly stated, "was laced with Darwinism from beginning to end. It's okay to teach Darwin, but you have to balance it with something else, such as creationism."

Buckingham watched the tape, which he must have known was coming, looked mildly irritated, and claimed that "when I was walking from my car to the building, here's this lady and here's a cameraman, and I had on my mind all the newspaper articles saying we were talking about creationism, and I had it in my mind to make sure, make double sure, nobody talks about creationism, we're talking intelligent design. I had it on my mind, I was like a deer in the headlights of a car, and I misspoke. Pure and simple, I made a human mistake."

In a rare instance of pushing the limit, Harvey asked him a few more questions, and then when Buckingham again pulled the "deer in the headlights" defense, Harvey asked for the tape to be played again. He then said, "You didn't look like you were very pressured to me. Is there something in that tape that suggests to you that you were feeling pressured at the time?"

"I can't help how it looks," said Buckingham. "I'm telling you I felt pressured at the time."

As he said, he was only human.

• • • •

Exactly how human he was I would not find out until much later, when I spoke to him in preparation for an on-camera interview.

Contrary to popular opinion, I don't think Buckingham was a "bad man" or, in all probability, all that much of "a liar," at least not in his own eyes. In fact, I think Buckingham was "a good man," and indeed this definition of himself was in a way his problem. When I spoke to him I actually rather liked him.

He was provocative (to say the least), but the provocation was often enjoyed for its own sake in an almost childlike way. The French expression "epater le bourgeois"—to shock the bourgeois—was transmuted by Buckingham into "shock the liberals." This he achieved by lobbing grenades into the conversation at random, such as referring to the ACLU as "the American Communist Lawyers Union."

Describing his problems with OxyContin, he said, "it grabs ahold of you and sneaks up. When I first started taking it I went on kind of a low dose, like 40 mg. And eventually it increased to where I was taking like 120 mg twice a day, and then I had OxyIR, which is OxyContin Immediate Relief for breakthrough pain. And what happens is in that process you start to change but you don't realize it, it gets to be normal ... I would get hot flashes, I mean I would go out in a snow storm with my pants and no shirt on and just breathe in cold air because I would get so hot. I was high all the time. I didn't realize it, you know ... I couldn't drive, I would fall asleep. I would make myself a cup of tea, set in a chair, fall asleep, spill it all over my lap. I would start to cook something on the stove and forget about it ... it's synthetic heroin is what it's equivalent to—and it's a bear."

He turned into a loner. During this time, two of his favorite aunts and two of his favorite uncles, his dog, and both his mother and his father died. One day his neighbor's daughter hanged herself, and he was the first person to arrive.

"And the OxyContin kind of acted as a shield ... when I went off the OxyContin all this stuff came rushing in. It hit me all at one time."

In February 2004, before the worst of the school board meetings, he knew he had to do something. He went to the Caron Foundation and began to detox. He was there for only a week when a friend of his died and he thought he ought to go home.

"So I went home, and it was the worst mistake I could have made. I found out that the reason the program lasts for a month is you have to have some counseling to go along with this ... I wasn't detoxed the whole way. I went through pure hell. I saw spiders that weren't there." He laughed briefly and then continued, "I went into a terrible depression, and the longer it went the worse it got." Through all of the most contentious school board meetings, he was wracked with pain and in despair. "I was so low, I'm telling you I could have sat on a dime and my feet wouldn't have touched the floor."

One day in September, it got so bad that he went fishing and took his gun along, "and I didn't plan on coming home." Before he could kill himself—not by putting the gun in his mouth, he told me, because he had seen the results of that while investigating suicides as a policeman—he ran into his second-grade teacher from the two-room schoolhouse he attended as a child.

"I talked to her, not about what I was going through, but just talked to her, and I changed my mind."

In December 2004, around the time the lawsuit was filed, he checked back into the Caron Foundation. This is where he was when Thompson and company were searching for him. He finally got the help he needed. "And I stayed the whole time, and everybody says I'm a different person now."

He regretted his rudeness, particularly to Jeff Brown. "I love Jeff to death. I mean he's a character, but he's funny, but he's influenced by his wife. She went to a very liberal college, and I think I was the first one on the board that could out-debate her and she didn't like it, and she turned on me big time."

A different person, perhaps, but a different person with the same views. From the tone of our conversation, it seemed to me that if he went back on the school board, he'd do it all again—just a little more politely. Looking back on the affair, he thinks it got blown out of proportion because the teachers' contract was coming up, and they knew they wouldn't get anything much out of this board. This was a way to get those board members out.

He claimed he had not been present when the evolution mural was burned, but he still hated the idea of men coming from monkeys. "I don't think there's proof . . . and it goes against Genesis in the bible. You know, I don't know how teachers who have professed to me that they're Christians . . . can go to church, support the bible, and then go to school and support Darwin. It can't be both ways . . . that goes with people who believe the Constitution is a living, breathing document that changes with time, and there are people that believe the bible's the same way. And that's nuts . . . It doesn't live and breath and change, just like gay marriage: the bible says that's an abomination for a man to lay with a man, a woman to lay with a woman. You know, that's black and white. I don't see how people can go to church and support that; that's nuts." He was against abortion, but he had no opinion one way or the other about contraception.

As far as he was concerned, the country was founded by Christians under Christianity.

"They came here from Europe to get away from a situation where they had a government-supported denomination. Now I didn't say church, I said denomination, and the people here wanted the right to worship God the way they wanted to worship the Christian God, not have to be a Methodist or a Catholic because the state said so . . . You know, if you want to come here and you want to worship Buddha or Allah, or whatever, fine, do it, but don't tell me that we can't pray before we have a football game, or at commencement kids can't talk about God because that's what they believe in, because this is a Christian country. If you don't like it go someplace where you'll be happy. Even if you're born here, this is still a Christian country. If you're not happy in a Christian country, go somewhere else. The planes leave every day—get on one and go."

As for the war in Iraq, his views were typically diplomatic in tone.

"Well, you know," he laughed, I think in anticipatory pleasure at the shock he was about to cause, "if I was the president of the United States, I'd pull all my troops out of there . . . and I'd nuke that place. I would nuke it. There wouldn't be a Muslim threat anymore 'cause there wouldn't be any more Muslims over there . . . We got people being slaughtered over there. If they want to play that game they're gonna lose if I'm the president, guaranteed."

I suggested that some people might think this wasn't a very Christian attitude. After all, you'd be nuking—

"Hey, it's a war. It's a war attitude. They did it in Japan, they can do it now . . . You know in World War Two, there were a lot of civilians killed when Hiroshima, when the bomb was dropped there, but the war ended, didn't it? And it saved a million American lives, and we've got to think of our country first. You know we're over there trying to do something for those people, and they got a bunch of idiots running around with towels on their heads blowing our guys up, you know. That's my solution: pull our people out and nuke the place—it's all over."

Bill and I, you may be surprised to hear, had several things in common. Neither of us went to college, and both of us enjoyed making inflammatory comments. But it went deeper than this. One of my last questions was about his parents. Who were they, what did they do, what were they like?

His father, he told me, fought in Japan in the Second World War, then worked as a line foreman, "you know, laying the rails and stuff like that," then went into construction. Bill was fond of him and thought he was a good guy. But his mother "was nothing but a drunk . . . You know she shot my dog one day when I was in school? And told me it ran away. And let me go hunting for it for over three weeks 'till I gave up. And I never knew she did that until my dad told me right before he died." At some point, she "decided to trip the light fantastic, you know, went out and had an affair, and next thing you know I got a brother . . . my dad stayed for me and raised the kid like he was his own."

My mother was also an alcoholic, and she too had an affair and gave birth to a son whom my father raised as his own. Here the similarities end. Bill and I had reacted to our pain in opposite ways. I was, and remain to this day, agonizingly perplexed by my mother's inability to fight her addiction when it caused such pain to me and the rest of the family. I assumed—or tried to assume—that she loved me, but if she loved me, why did she not stop drinking?

But I never thought of her as being bad. Instead, I developed a philosophy that excused all moral deficiency by seeing it as part of a psychological failing beyond the power of its owner. I left school young, like Bill, and after many other jobs went into the film business—an ugly business, perhaps, but one that, for a writer at least, was defined by attempts at subtle understandings of human nature.

Bill's life had been harsher, and its rules were more harshly defined. He had worked in a brutal world where, both physically and spiritually, simplified notions of "good" and "bad" were essential tools of survival. Add to this that he was raised in the Catholic church, where ideas of good and bad are starkly drawn, and then went into a fundamentalist church where such lines are even more rigorously drawn, and it became hard to imagine how such a life could possibly have produced a nuanced view of anything.

I am not prepared to argue, though, that my philosophy has proved a better balm to psychological pain than Bill's, and I'm sure that his is more effective than mine in many practical ways.

The Reporters

FOR ALL MY warm feelings toward Bill Buckingham, my fellow provocateur on the other side, you could hardly have listened to him in court without realizing that he was angering the judge and harming his own cause. But if you looked toward the back of the court, you saw Alan Bonsell sitting next to one of the new school board members appointed in the wake of all the resignations, a local pastor who resembled a sleepy dog (sometimes a *sleeping* dog)—and both were grinning. Their smiles seemed to say "This is all so silly and irrelevant."

When Bonsell took the stand a few days later, his grin soon vanished.

Before that happened, though, the two reporters who were on the ground when the whole saga began were called to the stand to testify to the truthfulness of their articles.

York was unusual in having two daily newspapers, the *York Daily Record* and the *York Dispatch*—the former a morning paper, the latter an afternoon paper—plus the *York Sunday News.* The Harrisburg *Patriot-News* and the Harrisburg bureau of the *Philadelphia Inquirer* also provided extensive coverage of local events.

The two reporters who would testify were never in court except on the day of their testimony, but I became friendly with many of the other local reporters. They were a hard-working, eccentric group, who, though nominally in competition with one another, worked together in collaborative amity.

Lauri Lebo, of the *York Daily Record,* was the daughter of a staunch fundamentalist who owned the local Christian radio station. Her views, however, were no longer the same as his—a fact that caused some distress between them. She was an intelligent woman married to a local musician and home builder who collected beer cans from all over the world. (In fact, he had collected so many they had to be stored in a large house next to the smaller one where he and Lauri lived.)

Playing in the same left-of-center band as Jeff Lebo was Mike Argento, who wrote the humor column in the *Daily Record.* Argento took few notes, but several lawyers told me they read his column and found it not only hilarious but an astonishingly accurate synthesis of the legal proceedings. Lebo and Argento were joined in court by Michelle Starr, and a fourth reporter for the *Daily Record,* Teresa McMinn, covered related events outside of court.

Christina Kauffman from the *York Dispatch* was not so well supported. She worked alone but made up for this by having an indomitable charm that gained her access wherever she went. Kauffman—who, when she wanted to interview me, said, "So, listen, would you consider having lunch with a short chubby girl?"—played on an all-female football team. In spite of her exhaustion, she never lost her sense of humor or her cool. It was she who one day realized that she had her pants on back to front.

Amy Worden, an excellent and experienced political journalist, worked in Harrisburg for the *Philadelphia Inquirer* and was married to the editor of the *York Dispatch*.

Bill Sulon of the Harrisburg *Patriot-News* was almost invariably dressed in a tweed jacket, which belied his true nature: he was a fanatical Frank Zappa fan. He was also a very canny reporter and was, like me, I soon found out, a germophobe. We took to comparing notes on who we thought was sickening and the wisdom of proximity. Before long, we began to refer to each other as Dr. Sulon and Dr. Chapman.

. . . .

The first reporter to take the stand was Heidi Bernhard-Bubb, who worked for the *York Dispatch*. She was a stay-at-home mother of two and was only a part-time reporter. She was an attractive blonde-haired woman with a nose so perfect it was almost architectural.

She and Joe Maldonado, of the *York Daily Record*, were testifying for the plaintiffs, but out of order because of some legal wrangling, so Vic Walczak had Heidi on direct examination and led her through events.

She testified that everything she had written was true. She had taken notes. She had heard the word "creationism" used at school board meetings by Buckingham and Bonsell on several occasions, and accurately reported Buckingham's many inflammatory remarks.

She did not seem like a liar and, as will become apparent later, had no reason to lie. In fact, if she was going to lie, you might imagine she'd lie the other way.

When Ed White cross-examined her, he suggested that similarities between her reports of the school board meetings and those of Maldonado resulted from collusion. According to him, they sat next to each other and made up their stories. The reason for this, he suggested, was that both papers were clearly proevolution and she was therefore—excuse the pun—pandering to this bias. (In fact, although the editorials did tend to favor evolution, both local papers were owned by people more likely to sympathize with the school board and its aims.)

Using the aggressive Ed White in this context was one of many mistakes the defense made. Bernhard-Bubb was a hard-working young woman trying to juggle motherhood and work—work in this case being the thankless task of reporting usually dreary school board meetings. Lauri Lebo was on the verge of tears as Ed questioned Heidi's qualifications, methods, and motives.

Joe Maldonado, part-time freelancer for the *York Daily Record*, received even harsher treatment. In his thirties, half Hispanic, a tough-looking guy with a shaved head and a goatee, he was handsome and polite in an almost military fashion. Indeed, it was soon revealed that he had served in the Air Force for almost seven years. His answers were a clipped "Yes, sir" or "No, sir" or "That is correct." He owned and ran a sandwich shop called PBJ's in York's Central Market House.

In the morning, Ed White had taken hold of the microphone on the lectern to adjust it, and it had broken. It had now been fixed. As Ed approached the lectern Judge Jones said, "Mr. White, we'll ask you not to tear the microphone from its moorings before you commence your cross-examination." Ed smiled, readjusted the mike, and broke it again.

"Well, at least the reporters have their lead story," he said.

Jones responded tartly, "That remains to be seen."

Ed smiled briefly, went through his previously described preening ritual, and turned his cold, contemptuous eyes on Joe.

"Mr. Maldonado, your primary occupation is running the sandwich shop?"

Maldonado replied that it was a toss-up between the sandwich shop and his writing.

"You don't have any formal training, though, correct?"

"No, sir."

"And freelancing: I know you love to write, but it's also a way to supplement your income, correct?"

"That is correct."

"And depending on where the article appears in the paper determines the amount of money you're paid per article, right?"

"Yes."

"So a front-page story gets you about $65?"

"$67.50."

"And then if it runs on a cover of one of the sections, the local sections, it's about $60?"

"$62.50."

"And then just your average story is around $50, right?"

"Somewhere in that ball park, yes."

"And it is the editors who decide where in the newspaper your stories will run, correct?"

Objections were raised, and the line of questioning died.

But there was something so moving in this exchange—Maldonado having to defend himself for working a double shift—one at his sandwich shop, the other as a reporter who got paid $50 per article—that I decided to pay him a visit.

. . . .

The sandwich shop was in the old Central Market in York, one of those cavernous spaces given over to stands selling crafts and bric-a-brac. Joe's place was rightfully famous for its Mojo Chicken sandwiches with cream cheese and wing sauce.

Working with him that day was the younger of his two sons, fourteen-year-old Jaryid. His older son, Alex, was at Penn State studying meteorology, and there was a jar on the counter for his college fund. Next to this was a self-published book of poems written by Maldonado. Hanging above the counter were two American flags.

Jaryid had open heart surgery when he was seven months old, which caused some developmental delays. A couple of years ago, Maldonado and his wife, Julie, though appreciative of the teacher's efforts, could see he was suffering in regular school. "It was so much for him. It was just overwhelming to go from one subject to the other, and I never got the sense that he was mastering one lesson before he'd move on to the next one." So they took him out, and now Joe "supplemented his income" through reporting, which enabled him to devote every afternoon to educating his son.

The liberal (by implication) reporter had not only been in the military ("I'm proud to say I served my country") but had also spent his first year of higher education at Liberty University, founded by Baptist evangelist Jerry Falwell. Although I did not learn this from Maldonado, and would not learn it until many months later when I interviewed Beth Eveland, who went to high school with him, Joe had been so devout in high school that he had started its first-ever bible club.

Brought up as a Baptist, he remained one. The trial had forced him to think a lot about the issues. His faith was so deeply embedded in him, he told me, that it was very hard to lose God from the equation. To him, the big question was whether intelligent design was "ready for prime-time science." He spoke eloquently about this and mentioned Darwin spending two decades collecting evidence before presenting his theory.

Before I left, I bought a copy of his book of poems, which he inscribed for me: "And the Lord said, 'Let there be . . .' Where's the science in that? Joe."

Later that night, I opened it with some trepidation and discovered that Joe wrote beautiful poems full of yearning and eroticism and a keen sense of sin. I remembered a question Ed White asked Joe that seemed a little strange at the time: "And freelancing, I know you love to write, but it's also a way to supplement your income, correct?"

"I know you love to write . . ." Why had he said that? Had he perhaps read Joe's book of poems? Was he toying with the idea of reading a few in court?

My next stop was to visit Heidi Bernhard-Bubb. She lived in the upper apartment of a nice house in York and had two children, Ulysses and Bronwyn, both below school age. Here the "liberal reporter" was found to be a practicing Mormon.

While studying at Brigham Young University, she had been in a band that she described as being a little like Franz Ferdinand. When her son, in order to impress this guest, started to say "Fuckie, fuckie, fuckie, fuckie," she remained unruffled. She was intelligent, funny, and likeable, and she disagreed with her church on such issues as gay marriage.

Things were not what they seemed. Or perhaps, more accurately, only on the outer edges did you find the authentic clichés, and when you found them, if you were me, those that you hated often turned out to be more poignant than repellent.

Heather Geesey, however, the next witness, fell squarely into the repellent category without mitigation. I found her the most terrifying of all the witnesses.

Bonsell and His Trinity of Loyal Women

A WOMAN WHO SEEMED to think—against all evidence—that everything she did and said was astoundingly adorable and funny, Heather Geesey clearly relished being on the same team as "President Alan," as she referred to Bonsell, and grinned relentlessly throughout as though appearing on a quiz show.

Vic Walczak asked her if she supported teaching intelligent design.

"Yes, because it gave a balanced view of evolution."

"It presented an alternative theory?"

"Yes."

"And the policy talks about gaps and problems with evolution?"

"Yes."

"You don't know what those gaps and problems refer to, do you?"

"No."

"But it's good to teach about those gaps and problems?"

"That's our mission statement, yes."

"But you have no idea what they are?"

"It's not my job, no."

"Is it fair to say that you didn't know much about intelligent design in October of 2004?"

"Yes."

"And you didn't know much about the book *Of Pandas and People* either, did you?"

"Correct."

"So you never participated in any discussions of the book?"

"No."

"And you made no effort independently to find out about the book?"

"No."

"And the administration had made copies of the book available to board members?"

"Yes."

"But you never read the book."

"No."

"And no one ever explained to you what intelligent design was about."

"No."

This went on for quite a while, Geesey grinning throughout as if her ignorance was just the cutest thing, as if a studio audience was out there somewhere wildly applauding in the dark, until finally, still with a blazing smile, she stated that she had relied on curriculum committee member Bill Buckingham to make the decision.

"And do you know whether Mr. Buckingham has a background in science?"

"No, I do not."

"Do you know that in fact he doesn't have a background in science?"

"I don't know. He's law enforcement, so I would assume he had to take something along the way."

Here it was in its most naked form: an auto repairman appointed an ex-cop and corrections officer, a biblical literalist without a shred of knowledge, to decide which books the kids should learn from—and a woman who had no curiosity about anything, even her own most deeply held beliefs, weighed in in support.

And, unless one doubted two seemingly decent professional reporters and a host of other witnesses, she would happily lie. As Geesey's testimony denying that the word "creationism" had been used at school board meetings became increasingly absurd, another side of Judge Jones became apparent. He turned and stared at her in evident exasperation.

As previously mentioned, he was determined never to be cruel, and his manners throughout were impeccable, but Geesey seemed to test the limits of his patience. In her deposition, Geesey had been asked whether she remembered when the phrase "intelligent design" was first used at school board meetings. She could not. Now, however, she was very clear it was in June, right at the start of the debacle. Perhaps sensing trouble, Patrick Gillen had asked her in his direct examination whether there was anything that had come to her attention in her preparation for this day that enabled her to "date with somewhat more precision" when she'd first heard the term. Geesey explained that it was by reading Beth Eveland's letter to the newspaper and her own response.

As she was about to leave the witness box Judge Jones stopped her.

"I have a question before you step down, Mrs. Geesey, because I'm confused."

"So am I," responded Geesey in typically perky fashion.

"Well," said Jones, with more than a hint of irritability, "it's more important that I'm not confused than that you're not confused." He wanted to know "what, in either of those letters, or both, leads you to believe that intelligent design was discussed at the June meeting?"

"I just—"

"What? Point me to what in the letter—not generally, but specifically."

"That I thought—"

"I asked you that question because I don't see the words 'intelligent design.'"

Geesey stumbled on, and eventually the judge gave up. Gillen asked if he could question the witness again under redirect. He obviously wanted to try and clean things up. By doing so he gave Walczak another chance to cross-examine her, which he did, pointing out that she'd been shown her letter at her deposition, and in fact been questioned on it quite rigorously, and it had not jogged her memory then. When Gillen asked for yet another opportunity to question Geesey, Jones said, "I don't know what you could possibly hope to achieve, but I'm going to give you one question and one only."

. . . .

The other two women who supported the school board's lust for *Pandas* were Sheila Harkins and Jane Cleaver. Cleaver was an old-time holy roller, so her motives for doing so could be assumed. Why Harkins supported it remained a mystery. She was a Quaker, and the Quakers had no problem with evolution.

Sheila came to court most days and sat in the back, watching. She was a short woman with short blonde hair, middle-aged but girlish. If Eugenie Scott was the class geek, Sheila had probably wanted to be a cheerleader but had not quite made the cut. She was perky like Geesey, but more middle class. Polite, even friendly at times, she was also defensive and irritable: What was all this fuss about? I mean, really!

As the time for her to take the stand approached, a persistent sniffle became a deluge. She claimed it was allergies. I usually sat at the end of the jury box, as close to the witnesses as I could get so I could look into their eyes. My fellow germophobe, Bill Sulon of the *Patriot-News*, tended to sit a little farther away, or even take shelter from potential germs in an unused room two floors down into which the uninfected audio was sent.

Both of us doubted Harkins's self-diagnosis of noninfectious allergies and were concerned about contagion. He considered outright flight to the lower room. I thought about finding a place in the jury box at farther remove from her legerdemain of tissues that fluttered through her hands and in and out of pockets and purses with extraordinary dexterity.

"I'm seeing no improvement in Harkins, Dr. Chapman," Sulon said to me on the day of her testimony. "I'll be at the back."

I concurred with his diagnosis—no remedy was going to stem this ceaseless and indubitably infectious flow—but the woman seemed so completely detached from reality, even the supposed reality of her own faith, that I conjectured she might actually say something so odd it would, as she blew holes in her tissues, blow holes in the case, and I wanted to see it up close. I scurried into the jury box to grab a chair a little farther away from the witness stand than my usual perch.

To everyone's astonishment, however, Harkins handled herself well, sharing with Geesey a certain insinuating perkiness but bluntly holding her own without recourse to either lies or inflammatory statements. She left the stand rightly proud of herself.

Next up was Jane Cleaver, a fragile old sparrow who left the board to migrate to Florida and had now been called back to face the consequences of pecking at the Constitution.

Sweet-smiling Patrick Gillen had her on direct, while Thomas Schmidt, the kindest and least peppery of the Pepper Hamilton lawyers, was brought in to ensure that no bones got broken.

Jane, the former owner of a five-and-dime, seemed like a nice old dear, but she did not even know the name of the theory she supported, constantly referring to intelligent design as *intelligence* design.

I'll say no more.

• • • •

When Alan Bonsell's time came, he strolled up to the stand with a good-natured grin and a few pleasantries for us reporters.

He admitted to Gillen that his personal views about the universe were based on the first two chapters of Genesis, but at no time had he tried to get creationism put into the science class. He believed evolution should be taught, but "when they don't include, you know, problems with it, or gaps in a theory, I mean, and you teach it, it almost sounds like they're teaching it as fact."

When asked to come up with an example of why he was so dubious about evolution, he said he'd seen something on one of the science channels about how the Piltdown Man was "the find of all finds and this was, this proved the evolutionary theory, and I think from that point up to the 1950s that was viewed that way until they found out that it was a hoax or a fraud."

He had also "seen things on different subjects of how bears turn into whales, you know, this was a natural scientific theory which I just thought was absurd. There's also statistical things that I've read about, how the statistical probability of life happening by itself was basically impossible, I mean statistically." He was not asked about the statistical probability of God slapping it together in six days.

One of the mysteries in the case (aside from who created the universe) was who had anonymously donated the sixty copies of *Of Pandas and People* to the school library. At various times in the past, both Buckingham and Bonsell had said they did not know.

In his testimony, however, Buckingham admitted he had gone to his church and asked for donations in order to buy them. He then gave the money to Bonsell's father, who bought the books and gave them to the school.

Steve Harvey, who had the plum job of cross-examining Bonsell, now took him back to his deposition, in which he denied knowing who was involved in this donation. It soon became abundantly clear that Bonsell knew—and had known at his deposition—the exact provenance of the books. Because depositions are taken under oath, he had therefore lied under oath. The exact motivation for lying never became entirely clear to me, but whatever it was, this potential perjury obviously offended Judge Jones a great deal.

When Harvey finished his cross-examination, Jones asked to see Bonsell's deposition, specifically the section about the donation of the books. He then turned to Bonsell.

"The specific question was asked to you, sir: 'You have never spoken to . . . anybody else who was involved with the donation?' And your answer was, 'I don't know the other people.' That didn't say, who donated? That said, *who was involved in* the donation?"

"Okay. I'm sorry. What—"

"Why did you—I'm on page 16."

"Okay."

"Line 9. That didn't say, 'who donated,' that said, 'who was *involved* in the donation.' Now you tell me why you didn't say Mr. Buckingham's name."

Bonsell stumbled and prevaricated, and the judge became increasingly irritated. Why, furthermore, had Bonsell's father been used in the transaction?

No clear answers were forthcoming, and afterwards there was much speculation that Bonsell and Buckingham, who also seemed to have lied under oath about the same matter, might face perjury charges.

Bonsell was clearly rattled. He had come onto the stand for the early part of his testimony chewing gum. That disappeared soon after Harvey started in on him. Now the swagger and the backward tilt of the head were also gone. He walked humbly back to the pews looking chastened, but within an hour or so, the old pose was back along with the gum.

At the end of Bonsell's testimony I found myself depressed. He seemed in some ways a nice enough man, certainly not without charm. I am speaking now from pure intuition, but I wondered whether he was as comfortable with his faith as he wanted you to think. His defense of religion was impatient, irritable, as if he had made a deal with himself: faith was essential, so it was therefore pointless to think about it or discuss it, even when it asked him to do things that were not really in his nature.

. . . .

My mood changed completely when I went that night to a meet-the-candidates event for Dover Cares, the group now seeking election to the school board.

Moderated by science teacher Rob Eshbach's father, Pastor Warren Eshbach, it was everything you might hope for in American politics.

Patricia Dapp, fifty-six, worked for the Family Health Counsel of Central Pennsylvania. She had two children and was involved in the Junior League of Harrisburg, the March of Dimes, and PA Hunger Action Center. Terry Emig, fifty, was a bus driver and a security officer. He had two kids and was the president of the Optimists Club of West York and a church counsel representative. He sang in the choir and was also involved with the Dover Band Boosters Association. Larry Guerreri, fifty-seven, was a manager at York County Nursing Home. Phillip Herman, fifty-five, had a lawn care business. He had been a Cub Scout, a Boy Scout, and a troop leader. He too was a member of the Dover Band Boosters. Herbert Robinson "Rob" McIlvaine, Jr, sixty-two, had a BA in English literature and a master's in education. He was vice president of MANTEC, one of the state's industrial resource centers, which was trying to promote economic development in the area. He had been involved with the American Cancer Society and the NAACP. Judy McIlvaine, his wife, was also sixty-two. A graduate of Vassar, from which she'd received a bachelor's degree in history, she was a semiretired editor and writer, was involved with the local public broadcasting station, and worked in a volunteer organization that helped maintain the Appalachian Trial. Bryan Rehm, the science teacher, you already know. Finally, there was Bernadette Reinking, fifty-eight, who had been educated at Fitzgerald Mercy School of Nursing. She was a retired nurse and a member of St. Rose of Lima Church.

I had met Rob McIlvaine on the street outside Dover High while he was campaigning with a couple of the other men. He wore a tweed jacket and discussed the situation in a solicitous, patrician manner. As someone trying to persuade industry to come to the area, McIlvaine found the antiscience school board to be his worst nightmare. This was exactly the image of rural Dover that he did not want conveyed.

The candidates sat in a line of chairs in a low-ceilinged room attached to a small modern library. In spite of two TV cameras and representatives of national and international newspapers, the event retained a pleasurable fervor, whose origin remained entirely local and was directed entirely at Dover.

If anyone breached the restraints of good manners it was me. Had I not asked the following question, I am pretty sure no one else would have. Dover Cares claimed to be entirely bipartisan, united only in its desire to get rid of the old school board and replace it with a less divisive one, so I asked whether the candidates would mind telling me which of them was Democrat and which Republican.

There was a moment of silent hesitation. Then, starting at one end of the line and passing all along its length, one by one they told me.

"Democrat." "Republican." "Democrat." "Republican." "Republican." "Democrat." "Democrat." "Republican . . ."

I can't remember what the exact breakdown was, but you could surely get no closer to bipartisanship. There was another moment of silence. Given the domineering, rancorous partisan rhetoric emanating from Washington, I don't think I was alone in feeling an unaccustomed pang of hope.

The Two Scotts

THE DEFENSE'S LAST expert witness, Scott Minnich, took the stand on the last day but one of the trial. If Michael Behe was like a professor out of a Tintin book, Scott Minnich was Tintin himself. A tenured professor of microbiology at the University of Idaho in Moscow (Moscow and Bethlehem in one trial!), Scott had volunteered for duty in Iraq to look for weapons of mass destruction: in his case, weapons of a biological nature.

Because his lab dealt with infectious diseases and worked with "organisms that were of concern," it was registered with the Centers for Disease Control and Prevention. Everyone in his lab had FBI clearance, including Minnich.

Like Michael Behe, Stephen Meyer, and William Dembski, he was a fellow of the Discovery Institute.

You could put me on the rack and apply the bastinado before I would ever join an organization that was, or had ever been, funded by a devotee of a crank like Rushdoony, with his murderous homophobia and his racism. Minnich, however, was proud of his association with the institute.

This was terrifying because this meant, presumably, that within Minnich's adventurous and wholesome Tintin heart there lurked accord with the messianic fervor expressed in The Wedge document:

> The social consequences of materialism have been devastating ... in order to defeat materialism, we must cut it off at its source ... Design theory promises to reverse the stifling dominance of the materialist worldview, and to replace it with a science consonant with Christian and theistic convictions.

"Devastating," "defeat," "cut it off at its source," "stifling dominance." To me this is the hysterical language of bin Laden, and here it was flowing down in Christian form upon a man who now told us he worked with "*Yersinia pestis,* which causes the bubonic plague ... an organism that is estimated to have killed two hundred million people in recorded history."

Minnich was a good-looking, all-American type with sandy hair, dressed in a blue blazer and tan slacks. You could cross him on the sidewalk and never know...

Yank that clearance!

To his credit, however, he did have a sense of humor, and it didn't seem to be the kind that would evolve into anything too twisted. Given that he and Behe were both biologists concentrating on the micro aspects of the craft, he didn't, as he himself admitted, have much to add to what Behe had already so comprehensively conveyed. When the ghastly specter of the bacterial flagellum once again reared up in court, the spectators groaned, and Judge Jones muttered amiably, "We've seen that."

Scott smiled ruefully and produced a familiar but apposite joke: "I feel like Zsa Zsa's fifth husband; you know? . . . I know what to do but I just can't make it exciting."

Robert Muise had, since Miller, found his scientific groove and seemed to understand the issues better than before—except, still, the primacy of "theory" over "fact"—and led Minnich through much complex testimony and the usual philosophical assertions about the movement. But by this time, we had heard so much other evidence to the contrary—including from the defense's other expert witness, Steve Fuller—that the sole impression left was that intelligent design, by its own admission, was, if it was science at all, not science in its infancy but a single determined but infertile sperm in search of a singularly elusive egg. It had found as little evidence for its validity as Scott had found evidence of biological weapons in Iraq, and given that it was unwilling to do its own research, it seemed unlikely to find any more in the future.

• • • •

On the last day of the trial, news cameras festooned the steps of the courthouse, and the court was jammed with spectators vibrating with excitement as Steve Harvey stood up to begin the case's final cross-examination.

As good as Robert Muise was in his relentless fashion, Harvey was better. In part, no doubt, he was stimulated by the thrilling atmosphere, but it was more than this. At some time between the testimonies of Michael Behe and Scott Minnich, Nick Matzke had found yet another detail with which to embarrass intelligent design, and Minnich was its unfortunate recipient.

Under Muise's expert examination, Minnich emphatically distanced himself from creationism and creation science. Like all other witnesses on his side, he denied that intelligent design had anything to do with religion, and asserted that it had no connection with, let alone genesis in, the creation science movement.

"I'm old enough that I was around during those debates, and I never participated because I don't agree with that approach. I don't think you mix religion with your science. I don't think you use Genesis as a filter of how you interpret your scientific data."

Minnich's interest in intelligent design had only solidified into something he was willing to advocate after he read *Darwin's Black Box,* and he had an almost reverent admiration for its author. He credited Behe with inventing the term "irreducible complexity" and bringing forth the bacterial flagellum as the great exemplar of the purposeful arrangement of parts.

Harvey asked Matthew McElvenny, the now weary but still faultless computer genius, to project a document on screen.

It was a page from the June 1994 issue of the *Creation Research Society Quarterly.* This was two years before Behe had published his book. On the page was a picture of the bacterial flagellum that was more or less indistinguishable from the ones Behe and Minnich had used to make their arguments. Most of the words and letters designating parts of the identically described "nanomachine" matched those in Minnich's diagram.

Within the text describing bacterial flagella were two illuminating passages.

> It is clear from the details of their operation that nothing about them works unless every one of their complexly fashioned and integrated components are in place.

> In terms of biophysical complexity, the bacterial rotor flagellum is without precedent in the living world . . . To evolutionists the system presents an enigma. To creationists it offers clear and compelling evidence of purposeful intelligent design.

Had Behe, the wondrous professorial conjurer of the highly scientific theory of "irreducible complexity," and the proponent of "the purposeful arrangement of parts," actually stolen his language and ideas from a Creationist tract?

The implication was certainly there, and not just in the above passages but in the slightly miffed and patronizing tone of Henry M. Morris, the founder of the Institute for Creation Research:

"We do appreciate the abilities and motives of Bill Dembski, Phil Johnson, and other key writers in the intelligent design movement. They think that if they can just get a wedge into the naturalistic mind set of the Darwinists, then later, the Biblical God can be suggested as the designer implicit in the concept." Morris went on to say, however, that this was not a new approach but rather one that employed "the same evidence and arguments used for years by scientific creationists ... These well-meaning folk did not really invent the idea of intelligent design, of course. Dembski often refers, for example, to the bacterial flagellum as a strong evidence for design, and indeed it is, but one of our scientists, the late Dr. Dick Bliss, was using this example in his talks on creation a generation ago."

Putting aside speculation on whether Dr. Bliss's mother was extraordinarily innocent or extraordinarily cruel in naming her son "Dick," the man did seem to have posthumously sodomized the Man from Bethlehem's claims both to originality and scientific purity.

I'm tempted to end this account now—it's always good to go out on a high, and you can't beat Bliss—but, though Bliss concluded Harvey's cross-examination, there were closing arguments and the judge's final remarks, all of which would happen after lunch.

As court recessed, Richard Thompson, the sword and shield for people of faith, got to his feet. With the final shots of the battle only an hour and a half away—and Patrick Gillen, who would fire them, apparently rallying from an ailment—the man who had valiantly proclaimed "the board stands fast, and the Thomas More Law Center is ready to represent them" began a lengthy and grandiloquent farewell.

He would not be in court that afternoon because he had "a long-standing commitment to be in the State of Oklahoma."

And off he went.

• • • •

At lunch, I stalked the plaintiffs' lawyers with my documentary crew and saw them huddled at a nearby coffee shop. Eric Rothschild, who was to make the closing arguments, looked nervous. After a while he walked away to go and think alone.

I bought a newspaper. The Department of Defense had just released the names of five more American soldiers killed in Iraq. This brought the total to 2,029. The number of Iraqi dead was not given. It was thought, however, to be well over 30,000.

After lunch, the court was again filled to capacity. Many of the lawyers' families were here, among them Rothschild's: Jill Rothschild had brought their two young children, Allison and Jake, who sat nervously in the pews as their father went to the lectern to make his closing arguments.

As most of what he said was a reiteration in synthesized form of all you have read so far, I won't transcribe it, but he ended with this:

> It's ironic that this case is being decided in Pennsylvania in a case brought by a plaintiff named Kitzmiller, a good Pennsylvania Dutch name. This colony was founded on religious liberty. For much of the 18th Century, Pennsylvania was the only place under British rule where Catholics could legally worship in public.
>
> In his declaration of rights, William Penn stated, "All men have a natural and indefeasible right to worship Almighty God according to the dictates of their own consciences. No man can of right be compelled to attend, erect, or support any place of worship or to maintain any ministry against his consent. No human authority can, in any case whatever, control or interfere with the rights of conscience, and no preference shall ever be given by law to any religious establishment or modes of worship."
>
> In defiance of these principles which have served this state and this country so well, this board imposed their religious views on the students in Dover High School and the Dover community. You have met the parents who have brought this lawsuit. The love and

> respect they have for their children spilled out of that witness stand and filled this courtroom.
>
> They don't need Alan Bonsell, William Buckingham, Heather Geesey, Jane Cleaver, and Sheila Harkins to teach their children right from wrong. They did not agree that this board could commandeer the religious education of their children and the Constitutions of this country, and this Commonwealth does not permit it.

He thanked the judge and sat down.

Patrick Gillen, like Eric Rothschild, covered all the important aspects of the case from his point of view and concluded by saying:

> In sum, Your Honor, I respectfully submit that the evidence of record shows that the plaintiffs have failed to prove that the primary purpose or primary effect of the reading of a four-paragraph statement to make the students aware of intelligent design, explaining that it's an explanation for the origins of life different from Darwin's theory, letting students know there are books in the library on this subject, does not, by any reasonable measure, threaten the harm which the Establishment Clause of the First Amendment to the United States Constitution prohibits, but, instead, the evidence shows that the defendants' policy has the primary purpose and primary effect of advancing science education by making students aware of a new scientific theory, one which Steve Fuller, accomplished by any man's measure, believes may well open a fascinating prospect to a new scientific paradigm.
>
> This is the very sort of legitimate educational goal which the United States Supreme Court acknowledged in *Edwards versus Aguillard*. For these reasons, I respectfully submit that this Court must deny the plaintiffs' request for relief and instead declare that Dover's curriculum is constitutional and enter a judgment dismissing the plaintiffs' claims with prejudice.

In typically gracious fashion, Judge Jones began his final remarks by thanking the security officers, the United States Marshals, the court reporters, and his own staff. He even had a kind word for the press.

Last he remarked, "Those of you who have sat through this trial ... have seen, by each and every one of the lawyers, some of the best presentations, some of the finest lawyering that you will ever have the privilege to see."

Perhaps emboldened by this, Patrick Gillen said, "Your Honor, I have one question, and that's this: By my reckoning, this is the fortieth day since the trial began, and tonight will be the fortieth night, and I would like to know if you did that on purpose?"

Jones smiled and instantly replied with the concluding line of the trial:

"Mr. Gillen, that is an interesting coincidence, but it was not by design."

. . . .

The judge's ruling would not come down for several weeks.

The plaintiffs moved to a local bar. Everyone was there: all the lawyers and their families, all the plaintiffs and their families, the legal assistants and their significant others. Even the estimable Wizard of Oz was sucking down beers with abandon.

Eric Rothschild, Steve Harvey, and Vic Walczak were buzzing with pride and optimism. Rothschild, elation compensating for the draining effects of his final effort, puffed on a cigar so large it looked as if it might topple him.

At times there were forty people here, laughing, slapping each other on the back, reliving moments from the trial, and reiterating their affection and admiration for one another.

By another coincidence, perhaps more fatefully determinative than that of the forty days and forty nights—perhaps a portent of a gift?—the last day of the trial fell on Tammy Kitzmiller's birthday. She was turning thirty-nine, so she and her daughters were celebrating twofold. Both girls had numerous ear piercings. Emboldened by a couple of beers, I asked Tammy whether she had any. Indeed she did: a discreet little belly-button ring.

When I looked up from examining this item, something caught my eye. There was an ill-lit restaurant across the street, in the window of which sat a small group of men hunched over a table. Something about them seemed familiar. I crossed the street to take a closer look.

The lawyers for the defense, absent the absent father, were dining alone, and from their lonely vantage point they could see the wild celebrations of the plaintiffs. These men had left their families—their gigantic families—to come and offer their skills free of charge to fellow Christians, but not one of their clients had thought fit to reward them with their company.

Exhausted and slightly drunk, I eventually returned to the Comfort Inn to pack. While covering the story I had gotten into the habit of asking anyone who looked interesting what they thought of the issues being discussed. Generally speaking, the answers were as limited and predictable as the temperature settings in my hotel room, with by far the largest group opting for the "Off" setting: "Don't know, don't care." I thought, however, that I should seek an opinion from one last person before leaving town.

Sitting outside the hotel was Scott Mehring of Mechanicsburg, Pennsylvania. Forty-seven years old, with a Rod Stewart haircut, Mehring was the one-time owner of a business that had something to do with performance cars. He wore a tight leather motorcycle jacket with no shirt visible underneath. He liked to party, he told me, and was ready to go out and party hard; but as he'd lost his license for various complicated reasons (which he told me about in great detail), he had no car, and his cab had not yet arrived, so he was happy to share his views with me until then.

I took out my recorder and recorded the following.

"If you go back to the Big Bang," he said, speaking rapidly, "the elements, I'm not sure exactly what they actually were, but whatever the elements were—the atom, the neutron, the proton neutron, whatever it was that created the Big Bang—*where* did that stuff come from? Somebody had to interject.

"Spontaneous generation is a dead theory—at one time they thought it was true—left a piece of meat on the ground maggots appeared: they thought the maggots came out of the meat, but you know they actually just came out to eat the food, so you can't say spontaneous generation created it . . . So there *had* to be some designer behind this.

"Now if you believe in physics, you got the eleventh dimension—it's a new theory, the eleventh dimension—and inside the eleventh dimension they say there's an infinite number of universes. So my take on that is that if you die on this earth, we just somehow hop over to the eleventh dimension, and hop from universe to universe to universe forever inside the eleventh dimension.

"So that means the bible could be right with everlasting life after we die. But, okay, the elements that started the Big Bang? If that was an intelligent designer, then you've got another complication. If there was like one dude somewhere at the very top that created everything? Well where did *he* come from? Who created him? And who created the God who created God? It gives me goosebumps. It's a loop, like in computer programming, it's an endless loop."

He paused and shook his head.

"If you think about this too much," he concluded as his cab arrived, "you can go insane."

The Genie Is Out of the Bottle

JUDGE JONES DROVE back to Pottsville to ponder his decision.

The plaintiffs returned to their lives in Dover. They were looking for another celebration, and they got it on November 9, when every single one of the Dover Cares candidates was elected to the school board, though some by the slimmest of margins.

Alan Bonsell and Sheila Harkins (the allegedly tight-fisted Quaker) received fewer votes than anyone. Only one vestige of the intelligent design movement remained: that paragon of intellectual curiosity, Heather Geesey, who was not up for reelection.

Among the new board members was Bryan Rehm. This put him in an odd position. As one of the plaintiffs he had sued the board. Now he was a member of that board. The plaintiffs had sought "declaratory and injunctive relief, nominal damages, costs, and attorney fees." The costs and attorney fees alone were now estimated to be approaching $2 million, so if Judge Jones ruled in favor of the plaintiffs, Bryan would be among those who had to find a way for Dover to pay the bill.

All this was dutifully recorded by the local reporters and picked up by the national press. For all their professionalism, however, the York and Harrisburg reporters I spoke to were all suffering post-trial blues. They had been part of a national debate such as had not been seen since the Scopes trial. With extraordinary generosity, they had welcomed, informed, and guided outsiders such as myself and had, I believe, enjoyed our company. Now we were gone, and they had to return to the sometimes dull reporting of local events.

Lauri Lebo went off on a trip to visit as many creationist museums as she could and then returned to work. Christina Kauffman's mother died, and she sank into a long depression. Both she and Lauri, whose fundamentalist father also died, had encounters with Reverend Groves, which increased their sense of loss. In Lauri's case, the preacher told her in no uncertain terms that unless she was born again she would not see her father in the great beyond.

My fellow hypochondriac, Bill Sulon, found it so hard to readjust to ordinary reporting that he soon decided to leave the business altogether. He became an investigator for the newly-formed Pennsylvania Gaming Control Board.

Steve Harvey went home to Philadelphia, to his corporate legal duties at Pepper Hamilton and his charitable work with the homeless. Like all the lawyers, he also had to reconnect with his family. He had, essentially, been away for six weeks: physically for most of it, emotionally and intellectually for all of it. His children were young, and even a week in the life of a toddler is a very long time.

Vic Walczak's family suffered the most. If the Pepper Hamilton lawyers had a day off they could shoot up to Philadelphia in under two hours. Walczak's drive home to see his wife and three children in Pittsburgh could take him over four hours, which, even putting aside the enormous amount of work required of him, often made it impossible.

Needless to say, apart from reacquainting himself with his children, Walczak had plenty to do at work. Among other things, he was involved in the defense of a group of thong-clad protesters arrested for recreating a scene from the Abu Ghraib prison scandal during a visit to the area by President Bush. In typically even-handed fashion, he also submitted an amicus brief in support of a preacher arrested for disorderly conduct after he had offended a lesbian by saying there was no such thing as a Christian lesbian. "While offensive, we think the remarks are protected by free speech."

Rothschild, the most restlessly ambitious of the trio—and, as the lead lawyer for the plaintiffs, the one who would be held most responsible for whatever transpired—was handed a distraction that combined both family and law.

His daughter, Allison, returned from elementary school one afternoon with a seemingly insurmountable legal problem. For a class project she had been asked to defend Cinderella's three ugly sisters in a mock trial. Rothschild threw himself into this, suggesting various legal strategies. Allison would argue that Disney, the means by which most kids received the story, had a bias against ugliness. Why was it *Beauty and the Beast*, not *Ugly and the Beast*; why *Sleeping Beauty*, not *Sleeping Ugly*? During cross-examination, Allison would get admissions from the other side that despite Cinderella's supposed bad treatment by her siblings, she had a roof over her head and she was always fed. She was, after all, in good enough condition to attract Prince Charming. When Allison argued her case, she won.

The defense lawyers returned to their families and their unique problems. During the trial, one of the younger lawyers, either Muise or Gillen, was heard discussing the logistical problems of having so many children. Even a minivan could not contain them all.

The Thomas More Law Center's crusading zeal continued unabated. Overturning *Roe v. Wade* was perhaps its most treasured aim, but other matters also required its attention. It tried to insert itself into a dispute about the possible removal of a large cross in San Diego. On its website it celebrated the fact that the new pope seemed to support intelligent design when he stated that God's love could be seen in the marvels of creation and quoted St. Basil the Great, a fourth-century saint, as saying that some people "fooled by the atheism they carry inside of them, imagine a universe free of direction and order as if at the mercy of chance."

Ed White got involved in the defense of a Detroit family who had been asked by a homeowners' association to remove a nativity scene from its lawn. The association soon backed down.

Richard Thompson, perhaps foreseeing defeat in *Kitzmiller v. Dover,* declaimed in a December 4 interview with the *Detroit News* that the outcome didn't really matter because whatever happened, intelligent design was now in the public mind.

"The genie is out of the bottle," he said.

It was indeed. And when Judge Jones issued his ruling on December 20, it received a monumental punch on the nose.

Jones's 139-page opinion briefly described the rise of fundamentalism in the nineteenth century, partially in response to Darwin; reviewed all the relevant precedents dealing with attempts to remove evolution from science classrooms, including the Scopes case; discussed the scientific validity and religious origins of intelligent design; and reached a stunningly emphatic conclusion.

Reviewing the testimony of not just the plaintiffs' experts, most specifically John Haught and Barbara Forrest, but also the statements of everyone in the intelligent design movement from Phillip Johnson, the founder of the movement, to Michael Behe, the defense's lead expert, he concluded that the purpose of the intelligent design movement was religious.

He then pondered the question of effect. Would an objective observer, watching the school board's efforts to "present" intelligent design while denigrating evolution, see these as "religious strategies that evolved from earlier forms of creationism"? He concluded that they would.

Perhaps most damning of all was that *Of Pandas and People* was published by a company that described itself as a Christian organization, and that the book's development from a creation science textbook to an intelligent design textbook could be clearly traced. He described as "astonishing" that a "purposeful change of words was effected without any corresponding change in content."

The statement, read in class before the students studied evolution, dishonestly portrayed the theory's status in the scientific community and wrongly suggested that it was atheistic.

Worse yet for the defense, Jones determined that intelligent design failed to qualify as science. It violated the centuries-old ground rules of science by "invoking and permitting supernatural causation." The argument of irreducible complexity, central to I.D., employed "the same flawed and illogical contrived dualism that doomed creation science." Its negative attacks on evolution were refuted by the scientific community, which had, in turn, rejected intelligent design.

As he recalled the details of the case, his tone became increasingly irate. The board's claimed reason for including intelligent design in the curriculum—solely because it was good science—was a "sham." He wrote:

> This case makes it abundantly clear that the Board's I.D. policy violates the Establishment Clause . . . Both Defendants and many of the leading proponents of I.D. make a bedrock assumption which is utterly false. Their presupposition is that evolutionary theory is antithetical to a belief in the existence of a supreme being and to religion in general. Repeatedly in this trial, Plaintiffs' scientific experts testified that the theory of evolution represents good science, is overwhelmingly accepted by the scientific community, and that it in no way conflicts with, nor does it deny, the existence of a divine creator.

To be sure, Darwin's theory of evolution is imperfect. However, the fact that a scientific theory cannot yet render an explanation on every point should not be used as a pretext to thrust an untestable alternative hypothesis grounded in religion into the science classroom or to misrepresent well-established scientific propositions.

The citizens of the Dover area were poorly served by the members of the Board who voted for the I.D. Policy. It is ironic that several of these individuals, who so staunchly and proudly touted their religious convictions in public, would time and again lie to cover their tracks and disguise the real purpose behind the I.D. Policy.

In conclusion, he said this:

Those who disagree with our holding will likely mark it as the product of an activist judge. If so, they will have erred as this is manifestly not an activist Court. Rather, this case came to us as the result of the activism of an ill-informed faction on a school board, aided by a national public interest law firm eager to find a constitutional test case on ID, who in combination drove the Board to adopt an imprudent and ultimately unconstitutional policy. The breathtaking inanity of the Board's decision is evident when considered against the factual backdrop which has now been fully revealed through this trial. The students, parents, and teachers of the Dover Area School District deserved better than to be dragged into this legal maelstrom, with its resulting utter waste of monetary and personal resources.

He ordered the defendants to stop requiring teachers to denigrate or disparage evolution or be forced to present the "religious alternative theory known as I.D," and he issued "a declaratory judgment that Plaintiffs' rights under the Constitutions of the United States and the Commonwealth of Pennsylvania have been violated by Defendants' actions."

The defendants would be subject to "liability with respect to injunctive and declaratory relief, but also for nominal damages and the reasonable value of Plaintiffs' attorneys' services and costs incurred in vindicating Plaintiffs' constitutional rights."

Although the legal costs would have to be paid by a school board that was now proevolution and had already removed intelligent design from the curriculum, the plaintiffs' lawyers believed it was essential to send a warning to other school boards that might be considering Dover-like strategies. The lawyers settled on an award of $2 million, which would be entered into the court record. This could be satisfied, however, by an actual payment of $1 million. Pepper Hamilton relinquished its lawyers' fees and took only out-of-pocket expenses. By August 31, 2006, the bill was paid.

• • • •

I went to Pennsylvania with a strong prejudice against Republicans, but discovered to my surprise that many of the people I most liked and respected were Republican. At one time or another I spoke with every plaintiff and always asked them at the end for their party affiliations. Their responses were completely unpredictable. If one can extrapolate from my experiences, there must be many Republicans who are not happy with the fundamentalist tenor of the current administration.

Most surprising of all was the definitive nature of Judge Jones's ruling. He was, after all, not just a Republican but a George W. Bush appointee. I decided to go up to Pottsville to get to know him better.

Pottsville is famous as the birthplace of John O'Hara, who immortalized it in his book *Appointment in Samarra,* in which he called it Gibbsville. O'Hara portrayed a town where manual labor of the hardest kind had thrown up a middle class with airs and graces, dinner parties, expensive cars, and highball-fuelled bacchanalia up at the country club.

By the time I went up there, those days were long gone. The main street was lined with shops, but few seemed occupied. Some were clearly defunct, abandoned with their merchandise still in the windows, muted by decades of sunlight and dust. Some remained defiantly alive. The most compelling of these was a shop selling wedding dresses and prom gowns of such garish hues (solar orange, blindingly assertive lilac, a pink so bold it made you blush) that no matter where you stood on the street your eye was drawn to it as to a virgin's shriek of laughter in a hospice.

This is not to say that Pottsville was ugly. On the contrary, from its small clapboard row houses to the county courthouse looming on the hill, it had a real beauty, but it was the beauty of an abandoned movie set, of a memory or of a dream.

John Jones's grandfather, the son of hardscrabble Welsh immigrants who came over in the 1870s, was a "breaker boy" in a local colliery, standing beside a conveyor belt as the coal went by and picking out the good lumps of coal from the bad, then breaking them into manageable pieces. Eventually, he took correspondence courses in engineering, bought his own stake in a mine, and soon bought three farms, where he raised Guernsey cows. Before long, he was part of the country club set.

After the Second World War, people started using oil and electricity for heat. The coal business went into steep decline, taking Pottsville with it. Only the fossil record on Main Street remained as evidence of its once prosperous past.

The Joneses, however, were an enterprising family. Using the heavy equipment from the mines, they carved out one of the first public golf courses in America. As Gary Player and Arnold Palmer popularized the game in the fifties and sixties, the social nature of the game changed, and people who could not afford to belong to country clubs wanted to play. Public golf courses became the Jones's primary business, and the family now owned five in Pennsylvania and New Jersey.

When I went to dinner at the judge's large and comfortable house on a hill at the edge of town, I was greeted by his wife, Beth. As humorous as her husband, she was lively, happy, intelligent, chirpily irreverent, and beautifully American, by which I mean she possessed that American quality of optimism and health which combines into a kind of radiance. Perhaps that is why the judges of the 1977 Schuylkill Winter Carnival Beauty Pageant crowned her Winter Carnival Queen.

She almost became Miss Pennsylvania.

A beauty queen who became a teacher, ceremony and substance—what better qualifications could you ask for in a first lady? Just as Judge Jones possessed a thoughtful decency absent in George W. Bush, so Beth possessed a kind of humanity absent in the steely-eyed Laura.

Before dinner, I asked Jones whether he had read any of the articles suggesting what a good president he would make and whether he ever considered such a thing. He replied, "I sometimes have a Walter Mittyish romantic notion of politics, but my rational side understands that it's not that way. I think it's debilitating, and I think it's family crushing. I think it's entirely disheartening, I think it can destroy good people, and so the tilt is away from that romantic 'Mr. Smith Goes To Washington' side."

I was disappointed. Having watched him in court for six weeks and having read his ruling in detail, I had come to believe Jones was a decent man and exactly the kind of person needed in American politics. After dinner, still thinking about his emphatic rejection of a political life, I probed a little more deeply. It soon became apparent as he described his experiences in politics, especially his congressional campaign, that what he hated most was being forced to clumsily define himself not as he was, not by what he truly thought and felt, but in terms that would get him elected.

The political process required that he become a breaker boy of himself, selecting and chopping up his subtle views and reducing them into rocks to throw at the opposition. If he wanted to win, he could no longer see any human issue as having two sides, as being as complex as humanity is, but instead had to transform everything into ugly and adversarial slogans.

He was a man who sought to weigh each case—each aspect of life—on its merits, on the evidence available, and reach conclusions only after compassionate reflection. Politics demanded that he walk around with an ax. He had been appointed to the bench for life. He enjoyed his work. He was not willing to exchange his gavel for a weapon.

In the spring, when *Time* magazine published its list of the hundred most influential people of the year, Judge Jones was among them.

This was not the only good thing to come out of Dover. Eric Rothschild and Steve Harvey from Pepper Hamilton, and Richard Katskee from Americans United for Separation of Church and State, were soon invited to speak all over the country, and Vic Walczak, though too busy with the ACLU to respond to all invitations, appeared at many such events. Eugenie Scott and Nick Matzke of the National Center for Science Education celebrated their victory and watched the Steve-o-meter broach 700 and soon rise to 750. They then returned their attention to creationist attacks in Kansas, Ohio, Alabama, and many other states.

As for the plaintiffs, their sense of vindication was palpable. At the risk of losing friends and enduring hatred from the community in which they lived, they had gathered together not to prevent a road from being built or to object to higher taxes, but to defend an idea. Some of them believed in God and some did not; some had been to college, others had not; but all of them were willing to stand up for a principle that the school board had attacked: the delicate principle of tolerance. They were good Americans, but even more they were good Pennsylvanians who had remained true to the founding spirit of their state.

Why then—to go back to the first paragraph of this book—am I for inviting creationism into the science classroom?

Revelation

INTELLIGENT DESIGN'S MENDACITY could be revealed by its own logic. Michael Behe believed that "the strong appearance of design in life is real and not just apparent." Explaining the irreducible complexity of the bacterial flagellum, he said "Most people who see this and have the function explained to them quickly realize that these parts are ordered for a purpose and, therefore, bespeak design."

Most people who examined the intelligent design movement saw that "the strong appearance of design" was indeed real and not just apparent, and that its parts had undeniably been "ordered for a purpose." In one fell swoop, Phillip Johnson had gathered his parts—Steve Meyer, William Dembski, and Michael Behe—and intelligently designed them into a "machine," the only function of which was to "reverse the stifling dominance of the materialist worldview and to replace it with a science consonant with Christian and theistic convictions."

It was appropriate that these men rode into battle on the back of a bacterium. Bacteria are unicellular microorganisms, many of which are parasites that cause disease. Those with flagella get around by flagellating the surrounding area with their tails. A better depiction of intelligent design could not be found. In *Kitzmiller v. Dover,* the bacterium was exposed to the antibiotic of reason, and will now either flagellate itself into oblivion or, more likely, mutate into a different form. In fact, one mutation is already at work. The Christian Camp and Conference Association boasts that 50 percent of its members now teach creationism. As an education director at the Timber-lee Christian Center in East Troy, Wisconsin, said, "The curriculum is designed to open their eyes so when they go back to school [and hear about evolution] they say, 'Oh, that sounds goofy!'"

Before *Kitzmiller v. Dover,* I assumed that antiscience fundamentalism had waned, but it had not. It had simply honed itself into a different shape. Now, with a presidential seal of approval stamped on its blade, it cut a gash through the country that revealed more about America than even the divisive issues of abortion or gay marriage.

The bisection was so well defined that you could almost imagine a Darwinian process at work, Descent with Modification, this being an explanation of how new species form: an animal population splits apart, and, under new circumstances, each group develops differently until eventually each becomes incapable of breeding with the other. I would never breed with a fundamentalist, and I'm sure none of them—no matter how much they relish breeding—would ever breed with me. When down in Dover, I often wondered, semihumorously, whether this would ultimately lead to two species of Americans: a huge, ignorant, but fertile fundamentalist species crowding out a dwindling species of smart but contraceptive-using intellectuals.

But if everything has a cause, there must be a cause of this. Why are fundamentalists suddenly so shrill and aggressive? Can they, like the canaries that miners took down into the mines to warn them of poisonous gases, actually be useful to us? Are they alarmed by something the rest of us should take more seriously?

The main proponents of intelligent design in Dover—Alan Bonsell, Bill Buckingham, and Sheila Harkins—were not awash in money flowing out of Silicon Valley, but nor were they about to lose their jobs to China. Alan Bonsell owned a small business that repaired cars and, like Harkins, bought and sold property. Until he retired, Bill Buckingham was a government employee in the expanding field of capturing and imprisoning criminals. They were middle or lower middle class. They had something to lose. Perhaps they had lost part of it already.

The Protestant work ethic upon which America was built always had financial prosperity as one of its chief goals. This was, however, inextricably linked to similarly positive spiritual aims. America was not just a resource to be mined; it was uniquely free and beautiful. It was blessed by God: if everyone prayed hard and worked hard, life in American would endlessly improve. So long as both elements of this dream remained believable, everyone was inspired to march optimistically forward on two strong legs.

Fifty years ago Buckingham, Bonsell, and Harkins could reasonably have expected to keep climbing the economic ladder and, on their death, leave their offspring on a higher rung. Now, with outsourcing of manufacturing jobs, porous borders, increased health care costs, a culture of debt, longer lives to strain the retirement system, out-of-control consumerism, and two expensive wars to pay for, the future looks less rosy. Suddenly they can no longer see the next rung up the ladder or even the one below. Beneath them is a dark void, while above is a massive gap, and then—way beyond their capacity to make the leap—a small extraordinarily wealthy group who lied to them when they promised to trickle some honey down the poles. Worse yet, many of these super-rich became so without seeming to work hard or make anything. From the pop star to the man who gambles on Wall Street or cooks the books at a corporation, the hero of the age is not one who struggles to assemble something that in turn will sustain a workforce, but the one who finds a way to sell an image or trick the system. How can anyone continue to believe in the Protestant work ethic when these contrary examples stare you in the face?

The Puritan acquired a credit card, bought himself a big TV, a computer, and a car, all made abroad, and hunkered down in his mortgaged home. But it's hard to relax when nothing you own actually belongs to you, when the Korean-made TV tells you that religious fanatics trying to destroy America will soon have access to nuclear and biological weapons, when the Japanese computer delivers pornography to your child, and when, as you drive to work in your imported car, a sidewise glance reveals a despairing America riddled with promiscuity, abortion, selfishness, greed, guns, and untreated insanity—a culture neither able to protect its weak through welfare nor inspiring enough to rouse them to employment, a nation that throws huge numbers in prison but cannot ultimately afford to keep them there, a society addicted to debt on one end and drugs at the other, with nothing at the center to hold it all together.

Is it surprising that, seeing this, a person looks elsewhere for comfort and meaning? And so fundamentalists transferred more of their weight to their other leg, the leg of religious faith. But this, too, had degenerated over time.

In most of Europe, members of the clergy were forced through college and developed some intellectual sophistication. In America, particularly among fundamentalist sects, preachers rose from the people and, without benefit of education, set up their stands, which frequently sold snake oil as well as salvation. Megachurches, TV begging, financial corruption and sexual scandal now lie at the end of this long tradition. It might be possible to hop around on such a gangrenous limb, but to imagine that you were purposefully marching forward to the triumphant beat of "Onward Christian Soldiers" required faith of truly delusional dimensions.

So, in spite of his considerable political power, the fundamentalist turned nasty. Thrown on his back—although he would not admit it—he required a target within easy kicking distance on which to exorcise his ire. Science was the perfect recipient. It had both economic and "spiritual" aspects, and—as if mirroring the rest of society—both had betrayed him.

Most people acquire knowledge not for its own sake but as a means of economic self-advancement. For years, science was sold as the best possible ticket to a golden secular future. Its seeming attacks on religion could be overlooked so long as it delivered. But with its technological progeny now being manufactured abroad, why not listen to the preachers who had despised science all along?

As for "enlightenment," here too science had failed to deliver. A hundred and fifty years ago, an educated man could comprehend the broad outlines of everything known about the universe. Now no one can. Paradoxically, the more science understands, the less accessible its truths become. The vast amounts of information being created by an increasing number of fractured scientific disciplines are beyond anyone's power to comprehend. The world is more baffling and mysterious than ever.

This is unsettling for everyone, but more so for the rural fundamentalist, who in his heart believes that someone else *does* understand it all—not necessarily someone richer, which he could tolerate—but someone smarter, someone who thinks he knows "the truth," someone "brainwashed in college." Intelligent design expertly pandered to this feeling of intellectual exclusion. How proud Bonsell and Buckingham were to finally own a scientific theory with which they could attack . . . science.

When I explained intelligent design's convoluted machinations to an investment banker, he shook his head and said, "Nothing destroys an economy faster than intellectual dishonesty." He might have added, "And the first to be destroyed will be those who are most ignorant because of it."

Unlike the miner's unwitting canary, fundamentalists are now willfully participating in their own destruction. Sublimating fear and resentment, they seek to deny their children the best chance of salvation in a modern world—an understanding of it—and instead retreat into dependence on the unchanging bible.

My favorite expert witness was John Haught, a Catholic theologian, and my favorite plaintiff was Beth Eveland, a Girl Scout leader on the road to Buddhism. If these two got together and formed a religion, I'd be there every Sunday. I would not be against *any* religion that had the humility to admit that it was just one of many equally valid "fairy tales of conscience" (to use philosopher George Santayana's definition of religion) or, in other words, "just a theory." But I fear all religions that claim sole ownership of absolute truth. With no evidence except that of ancient hearsay, they can only persuade through appeals to the most vulnerable aspects of the human mind, through menacing insistence on blind faith at the cost of reason, and ultimately through violence. Even if Christian Evangelicals in America rarely commit violence, their arrogant faith and missionary zeal certainly provoke it.

Faith in its widest sense—faith in anything for which there is no evidence—has reached epidemic proportions in America, and one has to wonder what the ultimate consequences of this will be. Books, movies, TV, music—all reflect the same desire to escape discomforting reality. Psychic detectives, people seeing dead people, heroes with superpowers, constant references to divine intervention in the form of angels and devils—even the secular hero, when his back is up against the wall, does not say to his aspiring acolyte "Have reason" but instead "Have faith."

The press, which should by definition be skeptical, is not. Flagrant hoaxes, pseudoscience, claims of paranormal powers, astrology, and even miracles are widely and uncritically, indeed approvingly, reported, while studies to the contrary—of which there are many—are frequently ignored.

As it is considered impolite to talk about faith, let alone examine it, and as neither art nor the press nor schools provide a critical sieve through which it has to pass, the message is clearly conveyed that faith is superior to reason and immune to it, that it exists in a higher realm. And if your religion tells you it is admirable to place faith above reason—and no one disagrees—why would you not apply the same principle, consciously or not, to all other aspects of life?

On one end of this spectrum is the man who knocks on wood for good luck. At the other is the suicide bomber who believes that if he blows himself up he will arrive in heaven. The immediate result of one act of faith is nothing, the immediate result of the other is horrific, but the logic behind each is identical: mystical forces guide the universe and can be accessed to our benefit in various ways. How can one not surmise that this kind of magical thinking lies behind many of the bad decisions people make?

At the start of the trial, Ken Miller said, "I think science might be the closest thing we have on this planet to a universal culture," and Barbara Forrest explained how the scientific method could be usefully applied to almost any aspect of life. The president of the United States thinks that "the jury is still out" on evolution and believes in "a divine plan that supersedes all human plans." He even claims access to it through his "personal relationship" with God. When Muslim jihadists say, "God is telling us to do this," Bush enthusiastically responds, "Yes, yes, he does that!"

The scientific method proposes a simple and humble means by which, through the examination of evidence, we distinguish what is true from what is not, what is the cause and what the cure. George W. Bush—in spite of almost no evidence that there were weapons of mass destruction in Iraq, nor any connection between Saddam Hussein and the events of 9/11—consulted with God and launched an invasion.

It is one thing if faith compels a few fundamentalists in Dover to attack evolution in the high school; it is another when a man who acts on similar beliefs becomes the most powerful man on earth.

Another Great Awakening is under way, which will, as always, insist that the intellect sink into a coma as great bloodshed ensues. A giant has woken from the Middle Ages and is thumping on the door, and Reason is too polite (and too afraid) to open the door and smash him in the face. Perhaps a better strategy, therefore, would be to use his massive weight and momentum against him. Open the door. Instead of pushing, pull. Let him, for example, tumble into science class—and then dissect him as he was dissected during *Kitzmiller v. Dover.*

Faith usually insists on protection from rational scrutiny. Here is an aspect of faith, creationism, that now *demands* examination. Fine. Let's throw away the frog that usually gets sacrificed to science and welcome this big fool who wants to sacrifice himself to both science *and* religion. If there is a better—not to mention safer—way to explore the limitations of faith through the revelatory methods of science, I cannot think of it. Creationism has no overtly violent implications, but its distorted logic is emblematic of other aspects of faiths that do.

Teachers would have to be retrained so they could present "The Dover Dissection" in a sensitive manner, but the benefits would be worth the effort. After all, a student currently in school will eventually become the president of the United States.

• • • •

By far the largest and most organized youth movement in this country is Evangelical. In a few years, a generation who believes that contraception is a mortal sin, that the Rapture lies around the corner if everyone will convert to Christianity, and that one of the central theories of science is "goofy" will start to vote. In a few more years its members will be running for office, and they will be running on an Evangelical agenda. While other kids are busy having sex and doing drugs, these ones are getting ready to take over America. If there is a "vast right-wing conspiracy," this is it.

If the "conspiracy" succeeds, a country founded by people escaping religious tyranny could soon be more tyrannized by religion than any of the countries from which the founders fled. Religion is already a more dominant political force in America than it is in any country in Europe. Apart from those run by Islamic fundamentalists, I can think of no other nation where organized religion already makes such frenzied attacks on women's rights, homosexuality, education, and science. In this sense, Bill Buckingham's suggestion—that anyone who does not like American religion should get on a plane and leave—is a warning that it might soon be necessary to jump back on the *Mayflower* and make a return journey for the same ostensible reasons as before.

But maybe there is some cause for optimism.

Eleven "ordinary" citizens in Dover bravely rejected the intrusion of fundamentalism into their lives and won. Maybe a new Age of Reason, a new Enlightenment, is not only possible but inevitable. After all, a situation now exists that has never existed before. We have reached a stage in our development where, to quote entomologist Edward O. Wilson, we are "the first species in the history of life to become a geophysical force." It may take us longer than forty days and forty nights, but through industrial pollution, the destruction of our rainforests, overfishing, overhunting—in short, overpopulation—we can destroy just about all life on earth. For the first time in our history, human beings have the power of God.

Darwin's theory of natural selection described a process acting on creatures without consciousness of their place in the universe, but we, as well as inventing the technology that can destroy us, have also evolved to a point where we can see that the selections we are now faced with are those we must make ourselves. The apocalypse is ours to choose or reject.

If there is something terrifying about this, there is also something inspiring in it. Perhaps horror at our power will force us to grow up. Maybe something more beautiful than religion will evolve to lead us forward: something wiser, a rephrasing of the Golden Rule perhaps, an Ecological Rule, an acknowledgement that everything we do in every aspect of our lives affects everyone and every thing, and that our highest duty is consideration of this simple scientific truth.

If Beth Eveland, a Girl Scout leader and the first plaintiff in *Kitzmiller v. Dover,* already holds this to be true, we can still have hope.